FEAR OF PHY

Lawrence M. Krauss is Ambrose Swasey Professor of Physics and Astronomy and Chairman of the Physics Department at Case Western Reserve University. The author of more than a hundred scientific publications as well as many popular articles on physics and astronomy, Krauss has made important contributions on issues ranging from the nature of exploding stars to the origin and nature of matter in the universe. He is also the author of *The Fifth Essence: The Search for Dark Matter in the Universe*, published by Vintage.

BY LAWRENCE M. KRAUSS

The Fifth Essence:
The Search For Dark Matter In The Universe

Fear Of Physics:
A Guide For The Perplexed

Lawrence M. Krauss

FEAR OF PHYSICS

A Guide for the Perplexed

VINTAGE

Published by Vintage 1996

2 4 6 8 10 9 7 5 3 1

First published in Great Britain by
Jonathan Cape Ltd, 1994

Vintage
Random House, 20 Vauxhall Bridge Road, London SW1V 2SA

Random House Australia (Pty) Limited
20 Alfred Street, Milsons Point, Sydney
New South Wales 2061, Australia

Random House New Zealand Limited
18 Poland Road, Glenfield,
Auckland 10, New Zealand

Random House South Africa (Pty) Limited
PO Box 337, Bergvlei, South Africa

Random House UK Limited Reg. No. 954009

A CIP catalogue record for this book
is available from the British Library

ISBN 0 09 930113 X

Papers used by Random House UK Ltd are natural, recyclable products made from wood grown in sustainable forests. The manufacturing processes conform to the environmental regulations of the country of origin

Printed and bound in Great Britain by
Cox & Wyman, Reading, Berkshire

To Kate, a constant source of inspiration

and

To Lilli, a constant source of joy

CONTENTS

PREFACE

WHEN SOMEONE at a party learns that I am a physicist, he or she immediately either (1) changes the subject, or (2) asks about the big bang, other universes, quarks, or one of the trilogy of recent "super" developments: superconductors, superstrings, or supercolliders. Even people who freely admit having avoided physics in high school and never looked back are still sometimes fascinated with the esoteric phenomena at the forefront of the field. Physics deals with many of the cosmic questions that, in one way or another, everyone has mused about. However, the field now often appears foreign and inaccessible, due in part to the fact that research at the forefront is at times far removed from everyday experience.

But there is a more basic obstacle that gets in the way of appreciating where modern physics is going. The way physicists approach problems, and the language they use, is also removed from the mainstream of modern-day activity for most people. Without a common *gestalt* to guide the observer, the menagerie of phenomena and concepts attached to modern developments remains disconnected and intimidating. So arises the fear of physics.

To break through this barrier and present modern physics as I understand it, it seems best to concentrate not on particular theories but rather on the tools that guide physicists in their work. If one wants to gain an appreciation for the current directions of modern physics, both as a human intellectual activity and as the basis for our modern picture of the universe, it is *much* easier and less intimidating to do so if you first have some notion of how the enterprise is carried out. What I want to present here, then, is not so much a *trail guide* through the modern physics jungle as a guide on *how to hike* in the first place: what equipment to bring, how to avoid cliffs and dead ends, what kind of trails are likely to be the most exciting, and how to get home safely.

Physicists themselves can follow modern developments only because they are largely based on the same handful of fundamental ideas that have been successfully applied to study the everyday world. Physical theory at present deals with phenomena occurring on scales of space and time varying by over sixty orders of magnitude—meaning that the ratio of the largest to the smallest is 1 followed by 60 zeros. Experiments cover a somewhat smaller range, but not much smaller. Yet amid this menagerie, any phenomenon described by one physicist is generally accessible to any other through the use of perhaps a dozen basic concepts. No other realm of human knowledge is either so extensive or so simply framed.

Partly for this reason, this book is short. The tools that guide physics are few in number, and while it may take an advanced degree to master them, it doesn't require a massive tome to elucidate them. So as you wander through each of the six chapters (once you BUY it, of course!), you will find a discussion of a key idea or theme that guides physicists in their search. To illustrate these ideas, I have chosen examples that run the gamut in physics, from the basics to the stuff the *New York Times* science writers will get confused about this week. The choice may seem at times eclectic. But, while concentrating at the beginning on what has guided physicists to get where we are, I will be concentrating by the end on what is guiding us to go where we are going.

Also, for this reason, I have taken the liberty to introduce concepts that are quite modern to illustrate a theme. Some readers may find these a pleasant relief from ideas they are already familiar with. Others may find them momentarily elusive. Some of these ideas, while fundamental, have never before been presented in the popular literature. No matter. You won't be tested. My intent is more to present the flavor of physics than it is to master its substance. I think it is insight, rather than a working knowledge, that is most useful and is needed by nonscientists in today's world, and so it is insight that I am aiming at.

Most important, there are subtle and wonderful connections between many of the short vignettes I shall present that run below the surface. It is these connections that form the fabric of physics. It is the joy of the theoretical physicist to discover them, and of the experimentalist to test their strength. In the end, they make physics accessible. If you get interested enough to crave more comprehensive discussions, there are lots of further resources.

Finally, I want to stress that physics is a *human* creative intellectual activity, like art and music. Physics has helped forge our cultural experience. I am not sure what will be most influential in the legacy we pass on, but I am sure that it is a grave mistake to ignore the cultural aspect of our scientific tradition. In the end, what science does is change the way we think about the world and our place within it. To be scientifically illiterate is to remain essentially uncultured. And the chief virtue of cultural activity—be it art, music, literature, or science—is the way it enriches our lives. Through it we can experience joy, excitement, beauty, mystery, adventure. The only thing that I think really differentiates science from the other things on this list is that the threshold is a little bit higher before the feedback starts. Indeed, a major justification for much of what we physicists do is the personal pleasure we get from doing physics. There is universal joy in making new connections. There is excitement and beauty in both the diversity of the physical world and the simplicity of its fundamental workings. So, with apologies to Erica Jong, this book is dedicated to the question: Is it possible for the average person to shed inhibitions, let go, and just enjoy the basic, simple pleasure of physics? I hope so.

ACKNOWLEDGMENTS

THIS BOOK would not have appeared at all, or at least would not have appeared in its present form, were it not for a number of people. First, Martin Kessler, president of Basic Books, conned me over breakfast almost ten years ago into turning my ideas about how physicists think about physics into what sounded like an ambitious book. Within a year we had signed a contract, which he kindly put on hold so that I could write another book for Basic on a subject that I then thought was more timely. My editor for that project, Richard Liebmann-Smith, became a good friend, and before he left Basic Books our conversations about this book helped refine my vision about what I wanted to accomplish.

Fear of Physics became something quite different from what I had originally envisaged. It became something I hoped my wife, Kate, would want to read. And, in fact, she provided constant input as I tested out my ideas and presentations on her. Indeed, the first chapter was not sent out to the publisher until it had earned her seal of approval for readability and interest. Finally, the new senior science editor at Basic Books, Susan Rabiner, played a vital role in bringing the present project to completion. It was she who convinced me that my new vision was workable and, more important, that Basic Books was prepared to produce and sell a book of this type. Once we had this settled and I had finally produced a chapter that conveyed what I wanted in the style I wanted, Susan was indefatigable. Her enthusiastic support for this book provided constant motivation. In particular, her scheduling of such things as the cover and the copyediting well in advance made my commitment seem ever more real and helped me complete the manuscript more or less on time—something new for me.

During the course of the writing, I have had the opportunity to discuss various ideas contained here with different people. As I have said, my wife often provided a filter through which things did not pass. I wish also to thank the numerous students I have taught over the years in physics courses for "nonscientists" who have helped me refine my thoughts by making it clear when something did not work. I fear I may have gained more from this process than they did. I also had the opportunity, through my work long ago at the Ontario Science Centre, to help build an appreciation of what nonphysicists might find comprehensible and what they might want to comprehend—often two different things. Finally, my teachers, and later my colleagues and collaborators, have influenced this work both directly and indirectly. There are too many individuals in this group to list by name. They know who they are, and I thank them. Next, as anyone who reads this book will quickly realize, Richard Feynman played an influential role in my thinking about a number of areas of physics, as I am sure he did for many physicists. I also want to thank Subir Sachdev for useful discussions that helped me refine my discussion of phase transitions in matter, and Martin White and Jules Coleman for reading the manuscript and providing comments.

Last but not least, I want to thank my daughter, Lilli, for lending me her computer during several periods when mine was broken. In a very real sense, this book would not have appeared now without her help. Both Lilli and Kate sacrificed precious time we could have spent together while I worked on this book, and I hope to make it up to them.

The initial mystery that attends any journey is: how did the traveller reach his starting point in the first place?

—*Louise Bogan,* Journey Around My Room

I

Process

1

Looking Where the Light Is

If the only tool you have is a hammer, you tend to treat everything as if it were a nail.

A PHYSICIST, an engineer, and a psychologist are called in as consultants to a dairy farm whose production has been below par. Each is given time to inspect the details of the operation before making a report.

The first to be called is the engineer, who states: "The size of the stalls for the cattle should be decreased. Efficiency could be improved if the cows were more closely packed, with a net allotment of 275 cubic feet per cow. Also, the diameter of the milking tubes should be increased by 4 percent to allow for a greater average flow rate during the milking periods."

The next to report is the psychologist, who proposes:

"The inside of the barn should be painted green. This is a more mellow color than brown and should help induce greater milk flow. Also, more trees should be planted in the fields to add diversity to the scenery for the cattle during grazing, to reduce boredom."

Finally, the physicist is called upon. He asks for a blackboard and then draws a circle. He begins: "Assume the cow is a sphere. . . ."

This old joke, if not very funny, does illustrate how—at least metaphorically—physicists picture the world. The set of tools physicists have to describe nature is limited. Most of the modern theories you read about began life as simple models by physicists who didn't know how else to start to solve a problem. These simple little models are usually based on simpler little models, and so on, because the class of things that we *do* know how to solve exactly can be counted on the fingers of one, maybe two, hands. For the most part, physicists follow the same guidelines that have helped keep Hollywood movie producers rich: If it works, exploit it. If it still works, copy it.

I like the cow joke because it provides an allegory for thinking simply about the world, and it allows me to jump right in to an idea that doesn't get written about too much, but that is essential for the everyday workings of science: *Before doing anything else, abstract out all irrelevant details!*

There are two operatives here: abstract and irrelevant. Getting rid of irrelevant details is the first step in building any model of the world, and we do it subconsciously from the moment we are born. Doing it consciously is another matter. Overcoming the natural desire *not* to throw out unnecessary information is probably the hardest and most important part of learning physics. In addition, what may be irrelevant in a given situation is not universal but depends in most cases on what interests you. This leads us to the second operative word: *abstraction.* Of all the abstract thinking required in physics, probably the most challenging lies in choosing how to approach a problem. The mere description of movement along a straight line—the first major development in modern physics—required enough abstraction that it largely eluded some pretty impressive intellects until Galileo, as I'll discuss later. For now, let's return to our physicist and his cow for an example of how useful even rather extreme abstraction can be.

Consider the following picture of a cow:

cow as a sphere

Now imagine a "supercow"—identical to a normal cow, except that all its dimensions are scaled up by a factor of two:

supercow

normal cow

What's the difference between these two cows? When we say one is twice as big as the other, what do we really mean? The supercow is twice the size, but is it twice as big? How much more does it weigh, for example? Well, if the cows are made of the same material, it is reasonable to expect that their weight will depend on the net amount of this material. The amount depends upon the *volume* of the cow. For a complicated shape, it may be difficult to estimate the volume, but for a sphere it's pretty easy. You may even remember from high school that if the radius is r, the volume is equal to $(4\pi/3)\ r^3$. But we don't have to know the exact volume of either cow here, just the ratio of their volumes. We can guess what this will be by recalling that volumes are quoted in cubic inches, cubic feet, cubic miles, and so on. The important word here is *cubic.* Thus, if I increase the linear dimensions of something by 2, its volume increases by the cube of 2, which is 2 × 2 × 2, or 8. So the supercow actually weighs 8 times as much as a normal cow. But what if I wanted to make clothes out of its hide? How much more hide would it yield than the normal cow? Well, the amount of hide increases as the surface area of the cow. If I increase the linear dimensions by 2, the surface area—measured in *square* inches, feet, miles, and so on—increases by the *square* of 2, or 4.

So a cow that is twice as "big" actually weighs 8 times as much and has 4 times as much skin holding it together. If you think about it, this means that the supercow has twice as much pressure pushing down on its skin as the normal cow does, due to its weight. If I keep increasing the size of a spherical cow, at a certain point the skin (or organs near the skin) will not have the strength to support this extra pressure and the cow will rupture! So there is a limit to how large even the most gifted rancher could breed his cows—not because of biology but because of the scaling laws of nature.

These scaling laws hold independent of the actual shape of the cow, so nothing is lost by imagining it as a simple shape like a sphere, for which everything can be calculated exactly. If I had tried to determine the volume of an irregularly shaped cow and to figure out how it changed as I doubled all the dimensions of the animal, I would have gotten the same result but it would have been harder. So for my purposes here, a cow is a sphere!

Now, we as we improve our approximation to the cow's shape, we can discover new scaling relations. For example, picture a cow slightly more realistically, as follows:

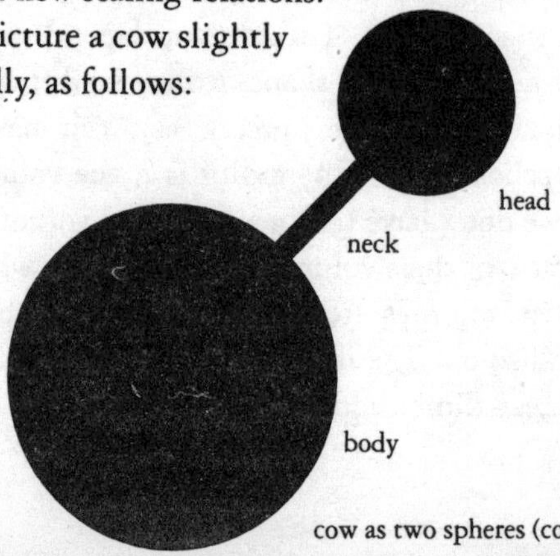

head

neck

body

cow as two spheres (connected by a rod)

The scaling arguments are still true not only for the whole cow but also for its individual parts. Thus, a supercow would now have a head 8 times more massive than that of a normal cow. Now consider the neck connecting the head to the body, represented here by a rod. The strength of this rod is proportional to its cross-sectional area (that is, a thicker rod will be stronger than a thinner rod made of the same material). A rod that is twice as thick has a cross-sectional area that is 4 times as large. So the weight of a supercow's head is 8 times as great as that of a normal cow, but its neck is only 4 times stronger. Relative to a normal cow, the neck is only half as effective in holding up the head. If we were to keep increasing the dimensions of our supercow, the bones in its neck would rapidly become unable to support its head. This explains why dinosaurs' heads had to be so small in proportion to their gigantic bodies, and why the animals with the largest heads (in proportion to their bodies), such as dolphins and whales, live in the water: Objects act as if they are lighter in the water, so less strength is needed to hold up the weight of the head.

Now we can understand why the physicist in the story did not recommend producing bigger cows as a way to alleviate the milk production problem! More important, even using his naive abstraction, we were able to deduce some general principles about scaling in nature. Since all the scaling principles are largely independent of shape, we can use the simplest shapes possible to understand them.

There's a lot more we could do with even this simple example, and I'll come back to it. First I want to return to Galileo. Foremost among his accomplishments was the precedent he created 400 years ago for abstracting out irrelevancies when he literally *created* modern science by describing motion.

One of the most obvious traits about the world, which makes a general description of motion apparently impossible, is that everything moves differently. A feather wafts gently down when loosened from a flying bird, but pigeon droppings fall like a rock unerringly on your windshield. Bowling balls rolled haphazardly by a three-year-old serendipitously make their way all the way down the alley, while a lawn mower won't move an inch on its own. Galileo recognized that this most obvious quality of the world is also its most irrelevant, at least as far as understanding motion is concerned. Marshall McLuhan may have said that the medium is the message, but Galileo had discovered much earlier that the medium only gets in the way. Philosophers before him had argued that a medium is essential to the very existence of motion, but Galileo stated cogently that the *essence* of motion could be understood only by removing the confusion introduced by the particular circumstances in which moving objects find themselves: "Have you not observed that two bodies which fall in water, one with a speed a hundred times as great as that of the other, will fall in air with speeds so nearly equal that one will not surpass the other by as much as one hundredth part? Thus, for example, an egg made of marble will descend in water one hundred times more rapidly than a hen's egg, while in air falling from a height of twenty cubits the one will fall short of the other by less than four finger-breadths."

Based on this argument, he claimed, rightly, that if we ignore the effect of the medium, all objects will fall exactly the same way. Moreover, he prepared for the onslaught of criticism from those who were not prepared for his abstraction by defining the very essence of *irrelevant:* "I trust you will not follow the example of many others who divert the discussion from its main intent and fasten upon some statement of mine which lacks a hair's-breadth of the truth and, under this hair, hide the fault of another which is as big as a ship's cable."[1]

This is exactly what he argued Aristotle had done by focusing not on the similarities in the motion of objects but on the differences that are attributable to the effect of a medium. In this sense, an "ideal" world in which there was no medium to get in the way was only a "hair's-breadth" away from the real one.

Once this profound abstraction had been made, the rest was literally straightforward: Galileo argued that objects moving freely, without being subject to any external force, will continue to move "straight and forward"—along a straight line at constant velocity—regardless of their previous motion.

Galileo arrived at this result by turning to examples in which the medium exerts little effect, such as ice underfoot, to argue that objects will naturally continue at a constant velocity, without slowing down, speeding up, or turning. What Aristotle had claimed was the natural state of motion—approaching a state of rest—is then seen to be merely a complication arising from the existence of an outside medium.

Why was this observation so important? It removed the distinction between objects that move at a constant velocity and objects that stand still. They are alike because both sets of objects will continue doing what they were doing unless they are acted upon by something. The only distinction between *constant* velocity and *zero* velocity is the magnitude of the velocity—zero is just one of an infinite number of possibilities. This observation allowed Galileo to remove what had been the focus of studies of motion—namely, the position of objects—and to shift that focus to *how* the position was changing, that is, to whether or not the velocity was constant. Once you recognize that a body unaffected by any force will move at a constant velocity, then it is a smaller leap (although one that required Isaac Newton's intellect to complete) to recognize that the effect of a force will be to *change* the velocity. The effect of a *constant* force will not be to change the *position* of an object by a constant amount, but rather its *velocity.* Similarly, a force that is changing will be reflected by a velocity whose *change* itself is changing! That is Newton's Law. With it, the motion of all objects under the sun can be understood, and the nature of all the forces in nature—those things that are behind all change in the universe—can be probed: Modern physics becomes possible. And none of this would have been arrived at if Galileo had not thrown out the unnecessary details in order to recognize that what really mattered was velocity, and whether or not it was constant.

Unfortunately, in trying to understand things exactly we often miss the important fundamentals and get hung up on side issues. If Galileo and Aristotle seem a little removed, here's an example that is closer to home. A relative of mine, along with several others—all college-educated individuals, one a high school physics teacher—invested over a million dollars in a project involving the development of a new engine whose only source of fuel was intended to be the Earth's gravitational field. Driven by dreams of solving the world's energy crisis, eliminating domestic dependence on foreign oil, and becoming fabulously wealthy, they let themselves be convinced that the machine could be perfected for just a little more money.

These people were not so naive as to believe you could get something for nothing. They did not think they were investing in a "perpetual motion" machine. They assumed that it was somehow extracting "energy" from the Earth's gravitational field. The device had so many gears, pulleys, and levers that the investors felt they could neither isolate the actual mechanism driving the machine nor attempt a detailed analysis of its engineering features. In actual demonstrations, once a brake on the machine was removed, the large flywheel began to rotate and appeared to gain speed briefly during the demonstration, and this seemed convincing.

In spite of the complexity of the machine's details, when these very details are ignored, the impossibility of the machine becomes manifest. Consider the configuration of the prototype I have drawn below at the start and the end of one complete cycle (when all the wheels have made one complete revolution):

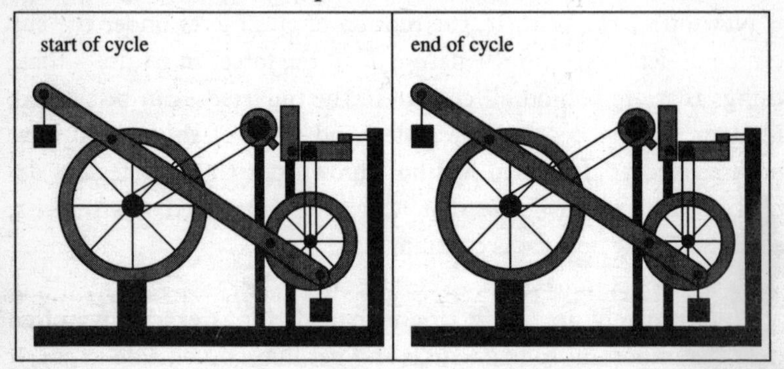

Every gear, every pulley, every nut, every bolt, is in exactly the same place! Nothing has shifted, nothing has "fallen," nothing has evaporated. If the large flywheel were standing still at the beginning of the cycle, how can it be moving at the end?

The problem with trying to perform an "engineering" analysis of the device is that if there are many components, it may be extremely difficult to determine exactly where and when the forces on any specific component cause the motion to cease. A "physics" analysis instead involves concentrating on the fundamentals and not the details. Put the whole thing in a black box, for example (a spherical one if you wish!), and consider only the simple requirement: If something, such as energy, is to be produced, it must come from inside; but if nothing changes inside, nothing can come out. If you try instead to keep track of everything, you can easily lose the forest for the trees.

How do you know in advance what is essential from what you can safely throw out? Often you don't. The only way to find out is to go ahead as best you can and see if the results make sense. In the words of Richard Feynman, "Damn the torpedoes, full speed ahead!"*

Consider, for example, trying to understand the structure of the sun. To produce the observed energy being emitted from the solar surface, the equivalent of a hundred billion hydrogen bombs must be exploding every second in its unfathomably hot, dense core! On the face of it, one could not imagine a more turbulent and complex environment. Luckily for the human species, the solar furnace has nevertheless been pretty consistent over the past few billion years, so it is reasonable to assume that things inside the sun are pretty much under control. The simplest alternative and, more important, perhaps the only one that lends itself even to the possibility of an analytical treatment, is to assume the inside of the sun is in "hydrostatic equilibrium." This means that the nuclear reactions going on

* Some attribute this to Admiral Dewey, but he was probably referring to something different.

inside the sun heat it up until just enough pressure is created to hold up the outside, which otherwise would collapse inward due to gravity. If the outside of the sun were to begin to collapse inward, the pressure and temperature inside the sun would increase, causing the nuclear reactions to happen more quickly, which in turn would cause the pressure to increase still more and push the outside back out. Similarly, if the sun were to expand in size, the core would get cooler, the nuclear reactions would proceed more slowly, the pressure would drop, and the outside would fall in a little. So the sun keeps burning at the same rate over long time intervals. In this sense, the sun works just like the piston in the engine of your car as you cruise along at a constant speed.

Even this explanation would be too difficult to handle numerically if we didn't make some further simplifications. First, *we assume the sun is a sphere*! Namely, we assume that the density of the sun changes in exactly the same way as we travel out from its center in any direction—we assume the density, pressure, and temperature are the same everywhere on the surface of any sphere inside the sun. Next, we assume that lots of other things that could dramatically complicate the dynamics of the sun, such as huge magnetic fields in its core, aren't there.

Unlike the assumption of hydrostatic equilibrium, these assumptions aren't made primarily on physical grounds. After all, we know from observation that the sun rotates, and this causes observable spatial variations as one moves around the solar surface. Similarly, the existence of sunspots tells us that the conditions on the solar surface are variable—its period of activity varies regularly on an eleven-year cycle at the surface. We ignore these complications both because for the most part they are too difficult to deal with, at least initially, and because it is quite plausible that the amount of solar rotation and the coupling between surface effects and the solar core are both small enough to ignore without throwing off our approximation.

So how good does this model of the sun work? Better than we probably have a right to expect. The size, surface temperature, brightness, and age of the sun can be fit with very high accuracy. More striking, perhaps, just as a crystal wineglass vibrates with sound waves when its rim is excited properly with your finger, just as the Earth vibrates with "seismic waves" due to the stresses released by an earthquake, the sun, too, vibrates with characteristic frequencies due to all the excitement happening inside. These vibrations cause motions on its surface that can be observed from Earth, and the frequencies of this motion can tell us a great deal about the solar interior, in the same manner that seismic waves can be used to probe the Earth's composition when searching for oil. What has become known as the Standard Solar Model—the model that incorporates all the approximations I just described—predicts more or less exactly the spectrum of oscillations at the solar surface that we observe.

It thus seems safe to imagine that the sun *really is* a simple sphere—that our approximate picture comes very close to the real thing. However, there is a problem. Besides producing an abundance of heat and light, the nuclear reactions going on inside the sun produce other things. Most important, they produce curious elementary particles, microscopic objects akin to particles such as electrons and quarks that make up atoms, called *neutrinos.* These particles have an important difference from the particles that make up ordinary matter. They interact so weakly with normal matter, in fact, that most neutrinos travel right through the Earth without ever knowing it is there. In the time it takes you to read this sentence, a thousand billion neutrinos originating in the fiery solar furnace have streamed through your body. (This is true day *or* night, since at night the neutrinos from the sun travel through the Earth to pierce you from below!) Since they were first proposed in the 1930s, neutrinos have played a very important role in our understanding of nature on its smallest scales. The neutrinos from the sun, however, have so far caused nothing but confusion.

The same Solar-Model calculations that so well predict all the other observable features of the sun should allow us to predict how many neutrinos of a given energy should arrive at the Earth's surface at any time. And while you might imagine that these elusive critters are impossible to detect, large underground experiments have been built with a great deal of ingenuity, patience, and high technology to do just that. The first of these, in a deep mine in South Dakota, involved a tank of 100,000 gallons of cleaning fluid, in which one atom of chlorine each day was predicted to be converted into an atom of argon by a chance interaction with a neutrino streaming from the sun. After twenty-five years, two different experiments sensitive to these highest-energy neutrinos from the sun have now reported their results. Both found fewer neutrinos than expected, between one-half and one-quarter of the predicted amount.

Your first reaction to this might be that it is much ado about nothing. To predict the results that closely could be viewed as a great success, since these predictions rely on the approximations of the sun's furnace which I have discussed. Indeed, many physicists took this as a sign that at least one of these approximations must be inappropriate. Others, most notably those involved with developing the Standard Solar Model, said this was extremely unlikely, given the excellent agreement with all other observables.

Of course, the only way to resolve such a debate is to conduct experiments that can test those aspects of the model that are insensitive to the bulk of the Solar Model approximations. Two such experiments are under way. As long as the nuclear reactions that power the sun are operating—and we know they are because the sun is shining—a certain number of low-energy neutrinos must be produced. The new experiments are sensitive to these neutrinos. The old ones weren't. *If* these new experiments were to confirm a deficit of these low-energy neutrinos, this would imply that something must be happening to the neutrinos on their way to Earth. This would in turn suggest that the resolution of the solar neutrino problem has nothing to do with inadequacies in our model of the solar interior, and everything to do with the properties of neutrinos. The properties would be likely to give important new insights into the branch of physics that deals with such elementary particles.

Unfortunately, the results so far have been inconclusive. It seems that we shall have to wait for a new generation of bigger or more sensitive detectors before this long-standing puzzle in physics will be resolved.

We can push still farther the approximation that the sun is a sphere, to probe still more of the universe. We can try to understand other stars, both bigger and smaller, younger and older, than the sun. In particular, the simple picture of hydrostatic equilibrium should give us a rough idea of the general behavior of stars over their entire lifetimes. For example, from the time stars first begin to form from collapsing gas until the nuclear reactions first turn on, the gas continues to get hotter and hotter. If the star is too small, the heat of the gas at prenuclear levels can provide sufficient pressure to support its mass. In this case, the star will never "turn on," and nuclear reactions will not begin. Jupiter is such an object, for example. For bigger clumps, however, the collapse continues until nuclear ignition begins and the heat released provides additional pressure which can stop further collapse and stabilize the system. Eventually, as the hydrogen fuel for the nuclear reactions begins to deplete, the slow inner collapse begins anew, until the core of the star is hot enough to burn the product of the first set of reactions, helium. For many stars this process continues, each time burning the product of the previous set of reactions, until the core of the star—now called a red or blue giant, because of the change of its surface color as the outside shell expands to great size at the same time as the inner core gets hotter and denser—is composed primarily of iron. Here the process must stop, one way or another, because iron cannot be used as nuclear fuel, due to the fact that the constituents of its nucleus—protons and neutrons—are so deeply bound together that they cannot give up any more binding energy by becoming part of a larger system. What happens at this point? One of two things. Either the star slowly dies down, like a match at the end of its stick or, for more massive stars, one of the most amazing events in the universe occurs: The star explodes!

An exploding star, or *supernova,* produces during its short fireworks display roughly as much light as a whole galaxy, over a hundred billion normal stars. It is hard to come to grips with the sheer power released during such an event. Just seconds before the final explosion begins, the star is calmly burning the last available bits of fuel, until the pressure generated by the star's last gasp is no longer sufficient to hold up its incredibly dense iron core, containing as much mass as our sun but compressed into a region the size of Earth—a million times smaller. In less than one second this entire mass collapses inward, in the process releasing huge amounts of energy. This collapse continues until the entire core is contained in a ball of roughly a 10-kilometer (6-mile) radius—about the size of New Haven, Connecticut. At this point, the matter is so dense that a teaspoonful would weigh thousands of tons. More important, the extremely dense atomic nuclei of the iron begin to crowd together, "touching," as it were. At this point, the material suddenly stiffens, and a whole new source of pressure, due to the interactions of these closely packed nuclei, becomes important. The collapse halts, the core "bounces," and a shock wave is driven outward, through the core, thousands of miles to the outer shell of the star, which is literally blown off and is visible to us as a supernova.

This picture of the collapse of the core of an exploding star was built up over decades of painstaking analytical and numerical work by teams of researchers after the first proposal by S. Chandrasekhar in 1939 that such an unfathomable scenario could take place. It is all an outgrowth of the simple ideas of hydrostatic equilibrium, which we believe governs the sun's structure. For fifty-odd years after first being proposed, the processes governing stellar collapse remained pure speculation. It had been centuries since the last supernova in our galaxy had been observed, and even then, all that could be seen were the outer fireworks, occurring far from where the real action was taking place, deep inside the star.

All of this changed on February 23, 1987. On that day a supernova was observed in the Large Magellanic Clouds, a small satellite system at the outer edge of our galaxy, about 150,000 light years away. This was the closest supernova to have been observed in the last four hundred years. It turns out that the visual fireworks associated with a supernova form just the tip of the iceberg. More than a thousand times as much energy is emitted, not in light, but—you guessed it—in almost invisible neutrinos. I say *almost* invisible because even though almost every neutrino emitted by the supernova could go through the Earth undetected, the laws of probability tell us that, although rarely, a neutrino will have a measurable interaction in a detector of smaller dimensions. In fact, one can estimate that at the moment that "neutrino blast" from the distant supernova passed across the Earth, one of every million or so people who happened to have their eyes closed at just the right time might have seen a flash induced by light produced when a neutrino bounced off an atom in their eyes.

Thankfully, however, we didn't have to depend on eyewitness accounts of this remarkable phenomenon. Two large detectors, each containing over 1,000 tons of water, located deep underground on opposite sides of the Earth had been outfitted with eyes for us. In each detector tank, thousands of photosensitive tubes lay ready in the darkness, and on February 23, in a ten-second period coincident in both detectors, nineteen separate neutrino-induced events were observed. Few as this may seem, it is almost precisely the number of events that one would predict to result from a supernova on the other side of our galaxy. Moreover, the timing and energy of the neutrinos were also in good agreement with predictions.

Whenever I think about this, I am still amazed. These neutrinos are emitted directly from the dense collapsing core, not from the surface of the star. They give us direct information about this crucial period of seconds associated with the catastrophic collapse of the core. And they tell us that the theory of stellar collapse—worked out in the absence of direct empirical measurement over thirty-odd years, and based on extrapolating to its extreme limits the same physics of hydrostatic equilibrium responsible for determining the sun's structure—is totally consistent with the data from the supernova. Confidence in our simple models led us to understand one of the most exotic processes in nature.

So, you may say, the approximations associated with modeling the sun are vindicated! Perhaps. Maybe the solar neutrino problem really is a problem in understanding neutrinos, not in understanding the sun. There remain some puzzles, however. If we use the same theory of stellar structure to predict how stars age, we can attempt to date not only our own sun (almost 5 billion years old) but also the oldest stars in our own galaxy. When stars in some isolated systems on the edges of our galaxy called globular clusters are examined and compared to the distribution of color, brightness, and other qualities predicted as a function of age, it is found that the oldest such systems are between 13 billion and 20 billion years old. At the same time, we can use the facts that our observed universe is expanding, and that this expansion is slowing down, to date the universe as a whole by measuring the rate of its expansion today. The slower the measured expansion of the universe today, the older the universe is. After sixty years of trying, we have been able to measure the expansion rate only to within a factor of 2, but even in the most optimistic case, the upper limit on the lifetime of the universe looks to be no greater than 14 billion years old. Thus, we are faced with the uncomfortable situation that our own galaxy appears to be perhaps older than the universe it is a part of! Clearly, one of the estimates may be somewhat skewed—whether it is the approximations associated with the theory of stellar structure, the simplest aspects of big bang cosmology, or the observations on which the stellar age estimates are based, remains to be seen.

Without approximation, we can do almost nothing. With it, we can make predictions that can be tested. When the predictions are wrong, we can then focus on the different aspects of the approximations we make, and in so doing we have learned almost everything we now know about the universe. In the words of James Clerk Maxwell, the most famous and successful theoretical physicist of the nineteenth century: "The chief merit of a theory is that it shall guide experiment, without impeding the progress of the true theory when it appears."[2]

Sometimes physicists simplify the world on the basis of sound intuition, but most often they do it because they have no other choice. There is a well-known allegory that physicists like to tell: If you are walking at night on a poorly lit street and you notice that your car keys are not in your pocket, where is the first place you look? Under the nearest streetlight, of course. Why? Not because you expect that you would necessarily lose your keys there, but rather because that is the only place you are likely to find them! So, too, much of physics is guided by looking where the light is.

Nature has so often been kind to us that we have come to take it sort of for granted. New problems are usually first approached using established tools, whether or not it is clear that they are appropriate, because it is all we can do at the time. If we are lucky, we can hope that even in gross approximation, some element of the essential physics has been captured. Physics is full of examples where looking where the light is has revealed far more than we had any right to expect. One of them took place shortly after the end of World War II, in a chain of events that carried with them elements of high drama and at the same time heralded the dawn of a new era in physics. The final result was the picture we now have of how physical theory evolves as we explore the universe on ever smaller or larger scales. This idea, which I never see discussed in popular literature, is fundamental to the way modern physics is done.

The war was over, physicists were once again trying to explore fundamental questions after years of war-related work, and the great revolutions of the twentieth century—relativity and quantum mechanics—had been completed. A new problem had arisen when physicists attempted to reconcile these two developments, both of which I shall describe in more detail later in the book. Quantum mechanics is based on the fact that at small scales, and for small times, not all quantities associated with the interactions of matter can be simultaneously measured. Thus, for example, the velocity of a particle and its position cannot be exactly determined at the same instant, no matter how good the measuring apparatus. Similarly, one cannot determine the energy of a particle exactly if one measures it over only a limited time interval. Relativity, on the other hand, stipulates that measurements of position, velocity, time, and energy are fundamentally tied together by new relationships that become more evident as the speed of light is approached. Deep inside of atoms, the motion of particles is sufficiently fast that the effects of relativity begin to show themselves, yet at the same time the scales are small enough so that the laws of quantum mechanics govern. The most remarkable consequence of the marriage of these two ideas is the prediction that for times that are sufficiently small so that it is impossible to measure accurately the energy contained in a certain volume, it is impossible to specify how many particles are moving around inside it. For example, consider the motion of an electron from the back of your TV tube to the front. (Electrons are microscopic charged particles, which, along with protons and neutrons, make up all atoms of ordinary matter. In metals, electrons can move about under the action of electric forces to produce currents. Such electrons are emitted by the metal tip of a heating element in the back of the TV and strike the screen at the front and cause it to shine, producing the picture you see.) The laws of quantum mechanics tell us that for any very short interval it is impossible to specify exactly which trajectory the electron takes, while at the same time attempting to measure its velocity. In this case, when relativity is incorporated into the picture, it suggests that if this is the case, during this short interval one cannot claim with certainty that there is only *one* electron traveling along. It is possible that spontaneously both *another* electron and its antiparticle with opposite charge, called a positron, can appear out of empty space and travel along with the electron for a short time before these two extra particles annihilate each other and disappear, leaving nothing behind but the original electron! The extra energy required to produce these two particles out of nothing for a short time is allowed for because the energy of the original electron moving along cannot be accurately measured over such short times, according to the laws of quantum mechanics.

After getting over the shock of such a possibility, you may think it sounds suspiciously close to trying to count the number of angels on the head of a pin. But there is one important difference. The electron-positron pair does not vanish without trace. Much like the famous Cheshire cat, it leaves behind a calling card. The presence of the pair can subtly alter the properties you attribute to the electron when you assume it is the only particle traveling during the whole affair.

By the 1930s it was recognized that such phenomena, including the very existence of antiparticles such as the positron, must occur as a consequence of the merging of quantum mechanics and relativity. The problem of how to incorporate such new possibilities into the calculation of physical quantities remained unsolved, however. The problem was that if you consider smaller and smaller scales, this kind of thing can keep recurring. For example, if you consider the motion of the original electron over an even shorter time, during which its energy is uncertain by even a larger amount, it becomes momentarily possible for not one electron-positron pair but two to hang around, and so on as one considers ever shorter intervals. In trying to keep track of all these objects, any estimate yielded an infinite result for physical quantities such as the measured electric charge of an electron. This, of course, is most unsatisfactory.

It was against this background that, in April 1947, a meeting was convened at an inn on Shelter Island, an isolated community off the eastern tip of Long Island, New York. Attending was an active group of theoretical and experimental physicists working on fundamental problems on the structure of matter. These included grand old men as well as Young Turks, most of whom had spent the war years working on the development of the atomic bomb. For many of these individuals, returning to pure research after years of directed war effort was not easy. Partly for this reason, the Shelter Island conference was convened to help identify the most important problems facing physics.

Things began auspiciously. The bus containing most of the participants was met by police troopers on motorcycles as they entered

Nassau County on western Long Island, and, to their surprise, they were escorted with wailing sirens across two counties to their destination. Later they found out that the police escort had been provided as thanks by individuals who had served in the Pacific during the war and had felt that their lives had been saved by these scientists who had developed the atomic bomb.

The preconference excitement was matched by sensational developments on the opening day of the meeting. Willis Lamb, an experimental atomic physicist, using microwave technology developed in association with war work on radar at Columbia University, presented an important result. Quantum mechanics, as one of its principal early successes, had allowed the calculation of the characteristic energies of electrons orbiting around the outside of atoms. However, Lamb's result implied that the energy levels of electrons in atoms were shifted slightly from those calculated in the quantum theory, which had been developed up to that time without explicitly incorporating relativity. This finding came to be known as the "Lamb shift." It was followed up by a report by the eminent experimental physicist I. I. Rabi on work of his, and also of P. Kusch, showing similar deviations in other observables in hydrogen and other atoms compared to the predictions of quantum mechanics. All three of these U.S. experimentalists—Lamb, Rabi, and Kusch—would later earn the Nobel Prize for their work.

The challenge was out. How could one explain such a shift, and how could one perform calculations that could somehow accommodate the momentary existence of a possibly infinite number of "virtual" electron-positron pairs, as they became called? The thought that the merging of relativity and quantum mechanics that caused the problem would also lead to an explanation was at that time merely a suspicion. The law of relativity complicated the calculations so much that up to that time no one had found a consistent way to perform them. The young rising stars of theoretical physics, Richard Feynman and Julian Schwinger, were both at the meeting. Each was independently developing, as was the Japanese physicist Sin-itiro Tomonaga, a calculational scheme for dealing with "quantum field theory," as the union of quantum mechanics and relativity became known. They hoped, and their hopes later proved correct, that these schemes would allow one safely to isolate, if not remove entirely, the effects of the virtual electron-positron pairs that appeared to plague the theory, while at the same time giving results that were consistent with relativity. By the time they were finished, they had established a new way of picturing elementary processes and demonstrated that the theory of electromagnetism could be consistently combined with quantum mechanics and relativity to form the most successful theoretical framework in all of physics—an accomplishment for which these three men deservedly shared the Nobel Prize almost 20 years later. But at the time of the meeting, no such scheme existed. How could one handle the interactions of electrons in atoms with the myriad of "virtual" electron-positron pairs that might spontaneously burp out of the vacuum in response to the fields and forces created by the electrons themselves?

Also attending the meeting was Hans Bethe, already an eminent theorist and one of the leaders in the atomic bomb project. Bethe would also go on to win the Nobel Prize for work demonstrating that nuclear reactions are indeed the power source of stars such as the sun. At the conference he was inspired by what he heard from both the experimentalists and the theorists to return to Cornell University to try to calculate the effect observed by Lamb. Five days after the meeting ended, he had produced a paper with a calculated result, which he claimed was in excellent agreement with the observed value for the Lamb shift. Bethe had always been known for his ability to perform complex calculations in longhand at the board or on paper flawlessly. Yet his remarkable calculation of the Lamb shift was not in any sense a completely self-consistent estimate based on sound fundamental principles of quantum mechanics and relativity. Instead, Bethe was interested in finding out whether current ideas were on the right track. Since the complete set of tools for dealing with the quantum theory including relativity were not yet at hand, he used tools that were.

He reasoned that if one could not handle the relativistic motion of the electron consistently, one could perform a "hybrid" calculation that incorporated explicitly the new physical *phenomena* made possible by relativity—such as the virtual electron-positron pairs—while still using equations for the motion of electrons based on the standard quantum mechanics of the 1920s and 1930s, in which the mathematical complexity of relativity was not explicitly incorporated. However, he found that the effects of virtual electron-positron pairs were then still unmanageable. How did he deal with that? Based on a suggestion he heard at the meeting, he did the calculation twice, once for the motion of an electron inside a hydrogen atom and once for a free electron, without the atom along with it. While the result in each case was mathematically intractable (due to the presence of the virtual particle pairs), he subtracted the two results. In this way he hoped that the difference between them—representing the shift in the energy for an electron located in an atom compared to that of a free electron not in an atom, precisely the effect observed by Lamb—would be tractable. Unfortunately, it wasn't. He next reasoned that this final intractable answer must be unphysical, so the only reasonable thing to do was to simplify it by some means, using one's physical intuition. He suggested that while relativity allowed exotic new processes due to the presence of virtual electron-positron pairs to affect the electron's state inside an atom, the effects of relativity could not be large for those processes involving many virtual electron-positron pairs whose total energy was much greater than energy corresponding to the rest mass of the electron itself.

I remind you that quantum mechanics allows such processes, in which many energetic virtual particles are present, to take place only over very small times because it is only over such small time intervals that the uncertainty in the measured total energy of the system becomes large. Bethe argued that if the theory including relativity was to be sensible, one should be able to ignore the effects of such exotic processes acting over very small time intervals. Thus he proposed to ignore them. His final calculation for the Lamb shift, in which only those processes involving virtual pairs whose total energy was less than or equal to the rest mass energy of the electron were considered, was mathematically tractable. Moreover, it agreed completely with the observations. At the time, there was no real justification for his approach, except that it allowed him to perform the calculation and it did what he assumed a sensible theory incorporating relativity should do.

The later work of Feynman, Schwinger, and Tomonaga would resolve the inconsistencies of Bethe's approach. Their results showed how in the complete theory, involving both quantum mechanics and relativity explicitly at every stage, the effects of energetic virtual particle-antiparticle pairs on measurable quantities in atoms would be vanishingly small. In this way, the final effects of incorporating virtual particles in the the theory would be manageable. The calculated result is today in such good agreement with the measured Lamb shift that this is one of the best agreements between theory and observation in physics! But Bethe's early hybrid approximation had confirmed what everyone already knew about him. He was, and is, a "physicist's physicist." He cleverly figured out how to use the available tools to get results. In the spirit of spherical cows, his boldness in ignoring extraneous details associated with processes involving virtual particles in quantum mechanics helped carry us to the threshold of modern research. It has become a central part of the way physicists approach the physics of elementary particles, a subject I shall return to in the final chapter of this book.

We have wandered here from cows to solar neutrinos, from exploding stars to Shelter Island. The common thread tying these things together is the thread that binds physicists of all kinds. On the surface, the world is a complicated place. Underneath, certain simple rules seem to operate. It is one of the goals of physics to uncover these rules. The only hope we have of doing so is to be willing to cut to the chase—to view cows as spheres, to put complicated machines inside black boxes, or to throw away an infinite number

Mathematicians deal with idealized structures, and they really don't care where, or whether, they might actually arise in nature. For them a pure number has its own reality. To a physicist, a pure number usually has no independent meaning at all.

Numbers in physics carry a lot of baggage because of their association with the measurement of physical quantities. And baggage, as anyone who travels knows, has a good side as well as a bad side. It may be difficult to pick up and tiresome to carry, but it secures our valuables and makes life a lot easier when we get to our destination. It may confine, but it also liberates. So, too, numbers and the mathematical relations among them confine us by fixing how we picture the world. But the baggage that numbers carry in physics is also an essential part of simplifying this picture. It liberates us by illuminating exactly what we can ignore and what we cannot.

Such a notion, of course, is in direct contradiction with the prevailing view that numbers and mathematical relations only complicate things and should be avoided at all costs, even in popular science books. Stephen Hawking even suggested, in *A Brief History of Time,* that each equation in a popular book cuts its sales by half. Given the choice of a quantitative explanation or a verbal one, most people would probably choose the latter. I think much of the cause for the common aversion to mathematics is sociological. Mathematical illiteracy is worn as a badge of honor—someone who can't balance his or her checkbook, for example, seems more human for this fault. But the deeper root, I think, is that people are somehow taught early on not to think about what numbers represent in the same way they think about what words represent. I was flabbergasted several years ago when teaching a physics course for nonscientists at Yale—a school known for literacy, if not numeracy—to discover that 35 percent of the students, many of them graduating seniors in history or American studies, did not know the population of the United States to within a factor of 10! Many thought the population was between 1 and 10 million—less than the population of New York City, located not even 100 miles away.

At first, I took this to be a sign of grave inadequacies in the social studies curriculum in our educational system. After all, the proximity of New York notwithstanding, this country would be a drastically different place if its population numbered only 1 million. I later came to realize that for most of these students, concepts such as 1 million or 100 million had no objective meaning. They had never learned to associate something containing a million things, like a mid-sized American city, with the number 1 million. Many people, for example, could not tell me the approximate distance across the United States in miles. Even this is too big a number to think about. Yet a short bit of rational deduction, such as combining an estimate of the mileage you can drive comfortably in a day on an interstate (about 500 miles) with an estimate of the number of days it would take to drive across the country (about 5–6 days) tells me this distance is closer to 2,500–3,000 miles than it is to, say, 10,000 miles.

Thinking about numbers in terms of what they represent takes most of the mystery out of the whole business. It is also what physicists specialize in. I don't want to pretend that mathematical thinking is something that everyone can feel comfortable with, or that there is some magic palliative cure for math anxiety. But it is not so difficult—often even amusing and, indeed, essential for understanding the way physicists think—to appreciate what numbers represent, to play with them in your head a little. At the very least, one should learn to appreciate the grand utility of numbers, even without necessarily being able to carry out detailed quantitative analyses oneself. In this chapter, I will briefly depart somewhat from Stephen Hawking's maxim (I hope, of course, that you and all the buying public will prove him wrong!) and show you how physicists approach numerical reasoning, in a way that should make clear why we want to use it and what we gain from the process. The central object lesson can be simply stated: We use numbers to make things never more difficult than they have to be.

In the first place, because physics deals with a wide variety of scales, very large or very small numbers can occur in even the simplest problems. The most difficult thing about dealing with such quantities, as anyone who has ever tried to multiply two 8-digit numbers will attest, is to account properly for all the digits. Unfortunately, however, this most difficult thing is often also the most important thing, because the number of digits determines the overall scale of a number. If you multiply 40 by 40, which is a better answer: 160 or 2,000? Neither is precise, but the latter is much closer to the actual answer of 1,600. If this were the pay you were receiving for 40 hours of work, getting the 16 right would not be much consolation for having lost over $1,400 by getting the magnitude wrong.

To help avoid such mistakes, physicists have invented a way to split numbers up into two pieces, one of which tells you immediately the overall scale or magnitude of the number—is it big or small?—to within a range of a factor of 10, while the other tells you the precise value within this range. Moreover, it is easier to specify the actual magnitude without having to display all the digits explicitly, in other words, without having to write a lot of zeros, as one would if one were writing the size of the visible universe in centimeters: about 1,000,000,000,000,000,000,000,000,000. Displayed this way, all we know is that the number is big!

Both of these goals are achieved through a way of writing numbers called *scientific notation.* (It should be called *sensible* notation.) Begin by writing 10^n to be the number 1 followed by *n* zeros, so that 100 is written as 10^2, for example, while 10^6 represents the number 1 followed by 6 zeros (1 million), and so on. The key to appreciating the size of such numbers is to remember that a number like 10^6 has one more zero, and therefore is 10 times bigger, than 10^5. For very small numbers, like the size of an atom in centimeters, about 0.000000001 cm, we can write 10^{-n} to represent the number 1 divided by 10^n, which is a number with a 1 in the nth place after the decimal point. Thus, one-tenth would be 10^{-1}, one billionth would be 10^{-9}, and so on.

This not only gets rid of the zeros but it achieves everything we want, because any number can be written simply as a number between 1 and 10 multiplied by a number consisting of 1 followed by *n* zeros. The number 100 is 10^2, while the number 135 is 1.35×10^2, for example. The beauty of this is that the second piece of a number written this way, called the *exponent,* or *power of ten,* tells us immediately the number of digits it has, or the "order of magnitude" of the number (thus 100 and 135 are the same order of magnitude), while the first piece tells you precisely what the value is within this range (that is, whether it is 100 or 135).

Since the most important thing about a number is probably its magnitude, it gives a better sense of the meaning of a large number, aside from the fact that it is less cumbersome, to write it in a form such as 1.45962×10^{13} rather than as 1,459,620,000,000, or one trillion, four hundred fifty nine billion, six hundred and twenty million. What may be more surprising is the claim I will shortly make that numbers that represent the physical world make sense *only* when written in scientific notation.

First, however, there is an immediate benefit of using scientific notation. It makes manipulating numbers much easier. For example, if you carry places correctly you find that $100 \times 100 = 10{,}000$. Writing this instead as $10^2 \times 10^2 = 10^{(2+2)} = 10^4$, multiplication turns into addition. Similarly, writing $1000 \div 10 = 100$ as $10^3 \div 10^1 = 10^{(3-1)} = 10^2$, division becomes as simple as subtraction. Using these rules for the powers of ten, the major headache—keeping track of the overall scale in a calculation—becomes trivial. The only thing you might need a calculator for is multiplying or dividing the first pieces of numbers written in scientific notation, the parts between 1 and 10. But even here things are simpler, since familiarity with multiplication tables up to 10×10 allows one to guess closely in advance what the result should be.

The point of this discussion is not to try to turn you into calculational whizzes. Rather, if simplifying the world means approximating it, then scientific notation makes possible one of the most important tools in all of physics: order-of-magnitude estimation. Thinking about numbers in the way scientific notation directs you to allows you quickly to estimate the answers to questions that would otherwise be largely intractable. And since it helps tremendously to know if you are on the right track when making your way through uncharted territory, being able to estimate the correct answer to any physical problem is very useful. It also saves a great deal of embarrassment. Apocryphal stories are told of Ph.D. students who presented theses with complex formulas meant to describe the universe, only to find during their thesis defense that plugging realistic numbers into the formulas shows the estimates to be ridiculously off.

Order-of-magnitude estimation opens up the world at your feet, as Enrico Fermi might have said. Fermi (1901–1954) was one of the last great physicists of this century equally skilled at experimental and theoretical physics. He was chosen to be in charge of the Manhattan Project, the secret U.S. wartime effort to develop a nuclear reactor and thereby demonstrate the feasibility of controlled nuclear *fission*—the splitting of atomic nuclei—in advance of building an atomic bomb. He was also the first physicist to propose a successful theory to describe the interactions that allow such processes to take place, for which he was awarded the Nobel Prize. He died at an untimely age of cancer, probably due to years of radiation exposure in the time before it was known how dangerous this could be. (Those of you who ever land at Logan Airport and get stuck in the massive traffic jams leading into the tunnel that takes you into Boston may amuse yourselves looking for a plaque dedicated to Fermi located at the base of a small overpass just before the tollbooths at the entrance to the tunnel. We name towns after presidents, and stadiums after sports heroes. It is telling that Fermi gets an overpass next to a tollbooth.)

I bring up Fermi because, as leader of a team of physicists working on the Manhattan Project in a basement lab under a football field at the University of Chicago, he used to help keep up morale by regularly offering challenges to the group. These were not really physics problems. Instead, Fermi announced that a good physicist should be able to answer any problem posed of him or her—not necessarily produce the right answer, but develop an algorithm that allows an order-of-magnitude estimate to be obtained based on things one either knows or can reliably estimate. For example, a question often asked on undergraduate physics quizzes is, How many piano tuners are there in Chicago at any given time?

Let me take you through the kind of reasoning Fermi might have expected. The key point is that if all you are interested in is getting the order of magnitude right, it is not so difficult. First, estimate the population of Chicago. About 5 million? How many people in an average household? About 4? Thus, there are about 1 million (10^6) households in Chicago. How many households have pianos? About 1 in 10? Thus, there are about 100,000 pianos in Chicago. Now, how many pianos does a piano tuner tune each year? If he is to make a living at it, he probably tunes at least two a day, five days a week, or ten a week. Working about fifty weeks a year makes 500 pianos. If each piano is tuned on average once a year, then 100,000 tunings per year are required, and if each tuner does 500, the number of tuners required is 100,000/500 = 200 (since $100{,}000/500 = 1/5 \times 10^5/10^2 = 1/5 \times 10^3 = 0.2 \times 10^3 = 2 \times 10^2$).

The point is not that there may or may not be exactly 200 piano tuners in Chicago. It is that this estimate, obtained quickly, tells us that we would be surprised to find less than about 100 or more than about 1,000 piano tuners. (I believe there are actually about 600.) When you think about the fact that before performing such an estimate, you probably had no idea what the range of the answer might be, the power of this technique becomes manifest.

Order-of-magnitude estimation can give you new insights about things you might never have thought you could expect to estimate or picture. Are there more specks of sand on a beach than there are stars in the sky? How many people on Earth sneeze each second? How long will it take wind and water to wear down Mount Everest? How many people in the world are . . . (*fill in your favorite possibility here*) as you read these words?

Equally important, perhaps, order-of-magnitude estimation gives you new insights about things you *should* understand. Humans can directly picture numbers up to somewhere between 6 and 12. If you see the 6 dots when you roll a die, you don't have to count them each time to know there are 6. You can picture the "whole" as distinct from the sum of its parts. If I gave you a die with 20 sides, however, it is unlikely that you could look at 20 dots and immediately comprehend them as the number 20. Even if they were arranged in regular patterns, you would still probably have to group the dots in your head, into, say, 4 groups of 5, before you could discern the total. That does not mean we cannot easily intuit what the number 20 *represents.* We are familiar with lots of things that are described by that number: the total number of our fingers and toes; perhaps the number of seconds it takes to leave the house and get into your car.

With truly big or small numbers, however, we have no independent way of comprehending what they represent without purposefully making estimates that can be attached to these numbers to give them meaning. A million may be the number of people who live in your city, or the number of seconds in about 10 days. A billion is close to the number of people who live in China. It is also the number of seconds in about 32 years. The more estimates you make of quantities that have these numbers attached to them, the better you can intuitively comprehend them. It can actually be fun. Estimate things you are curious about, or things that seem amusing: How many times will you hear your name called in your life? How much food do you eat, in pounds, in a decade? The pleasure you get from being able to tackle, step by step, what would be an insurmountably difficult problem to answer exactly can be addictive. I think this kind of "rush" provides much of the enjoyment physicists get from doing physics.

The virtue of scientific notation and order-of-magnitude estimation is even more direct for physicists. These allow the conceptual simplifications I discussed in the last chapter to become manifest. If we can understand the correct order of magnitude, we often understand most of what we need to know. That is not to say that getting all the factors of 2 and pi correct is not important. It is, and this provides the acid test that we know what we are talking about, because we can then compare predictions to observations with ever higher precision to test our ideas.

This leads me to the curious statement I made earlier, that numbers that represent the world make sense *only* when written in scientific notation. This is because numbers in physics generally refer to things that can be *measured.* If I measure the distance between the Earth and the sun, I could express that distance as 14,960,000,000,000 or 1.4960×10^{13} centimeters (cm). The choice between the two may seem primarily one of mathematical semantics and, indeed, to a mathematician these are two different but equivalent representations of a single identical number. But to a physicist, the first number not only means something very different from the second but it is highly suspect. You see, the first number suggests that the distance between the Earth and sun is 14,960,000,000,000 cm and not 14,959,790,562,739 cm or even 14,960,000,000,001 cm. It suggests that we know the distance between the Earth and sun to the nearest centimeter!

This is absurd because, among other things, the distance between Aspen, Colorado (at noon, Mountain Standard Time), and the sun, and New Haven, Connecticut (at noon, Eastern Standard Time), and the sun differ by 8,000 feet, or about 250,000 centimeters—the difference between the heights of Aspen and New Haven above sea level. Thus, we would have to specify where on Earth we made such a measurement to make it meaningful. Next, even if we specify that this distance is from the center of the Earth to the center of the sun (a reasonable choice), this implies that we can measure the size of the Earth and sun to the nearest centimeter, not to mention the distance between the two, to this precise accuracy. (If you think about any practical physical way in which you might go about measuring the distance from the Earth to the sun, you can convince yourself that a measurement with this kind of accuracy is unlikely, if not impossible.)

No: It is clear that when we write 14,960,000,000,000 cm, we are rounding off the distance to a tidy number. But with what accuracy do we really know the number? There is no such ambiguity when we write 1.4960×10^{13} cm, however. It tells us exactly how well we know this distance. Specifically, it tells us that the actual value lies somewhere between 1.49595×10^{13} cm and 1.49605×10^{13} cm. If we knew the distance with 10 times greater accuracy, we would write instead 1.49600×10^{13} cm.

Thus, there is a world of difference between 1.4960×10^{13} cm and 14,960,000,000,000 cm. More than a world, in fact, because if you think about the uncertainty implicit in the first number, it is about 0.0001×10^{13}, or 1 billion, centimeters—greater than the radius of the Earth!

This leads to an interesting question. Is this number accurate? While an uncertainty of 1 billion centimeters seems like an awful lot, compared to the Earth-to-sun distance it is small—less than 1 ten-thousandth of this distance, to be exact. This means that we know the Earth-to-sun distance to better than 1 part in 10,000. On a relative scale, this is highly accurate. It would be like measuring your own height to an accuracy of a tenth of a millimeter.

The beauty of writing down a number like 1.4960×10^{13} is that because the 10^{13} sets the "scale" of the number, you can immediately see how accurate it is. The more decimal places that are filled in, the higher the accuracy. In fact, when you think about it this way, numbers written in scientific notation tell you above all what you can *ignore!* The moment you see 10^{13} cm, you know that physical effects that might alter the result by centimeters, or even millions or billions of centimeters, are probably irrelevant. And as I stressed in the last chapter, knowing what to ignore is usually the most important thing of all.

I have so far ignored perhaps the most crucial fact that makes 1.49600×10^{13} cm a physical quantity and not a mathematical one. It is the "cm" tacked on the end. Without this attribute, we have no idea what kind of quantity it refers to. The "cm" tells us that this is a measurement of length. This specification is called the *dimension* of a quantity, and it is what connects numbers in physics with the real world of phenomena. Centimeters, inches, miles, light-years—all are measurements of distance and so carry the dimensions of length.

The thing probably most responsible for simplifying physics is a fascinating property of the world. There are only three kinds of fundamental dimensional quantities in nature: length, time, and mass.* *Everything, all physical quantities,* can be expressed in terms of some combination of these units. It doesn't matter whether you express velocity in miles/hour, meters/sec, furlongs/fortnight, they are all just different ways of writing length/time.

This has a remarkable implication. Because there are just three kinds of dimensional quantities, there are a limited number of independent combinations of these quantities you can devise. That means that every physical quantity is related to every other physical quantity in some simple way, and it strongly limits the number of different mathematical relations that are possible in physics. There is probably no more important tool used by physcists than the use of dimensions to characterize physical observables. It not only largely does away with the need to memorize equations but it underlies the way we picture the physical world. As I will argue, using dimensional analysis gives you a fundamental perspective of the world, which gives a sensible basis for interpreting the information obtained by your senses or by other measurements. It provides the ultimate approximation: When we picture things, we picture their dimensions.

When we earlier analyzed the scaling laws of spherical cows, we really worked with the interrelationship between their dimensions of length and mass. For example, what was important there was the relationship between length and volume and, more explicitly, the ratio of volumes of objects that were scaled up in size. Thinking about dimensions, we can go further and figure out how to estimate the volume itself of any object. Think of any system of units to de-

* One might choose to add electric charge to this list, but it is not necessary. It can be expressed in terms of the other three.

scribe the volume of something: cubic inches, cubic centimeters, cubic feet. The key word is *cubic.* These measurements all describe the same dimensions: length × length × length = length^3. Thus, it is a good bet that the volume of an object can be estimated by picking some characteristic length, call it *d,* and then cubing it, or taking d^3. This is usually good to within an order of magnitude. For example, the volume of a sphere I gave earlier can be rewritten as $\pi/6\ d^3 \approx [1/2]\ d^3$, where *d* is its diameter.

Here's another example: Which is correct: distance = velocity × time, or distance = velocity/time? Though the simplest kind of "dimensional analysis" can immediately give the correct answer, generation after generation of students taking physics insists on trying to memorize the formula, and they invariably get it wrong. The dimensions of velocity are length/time. The dimensions of distance are length. Therefore, if the left-hand side has dimensions of length, and velocity has dimensions of length/time, clearly you must multiply velocity by time in order for the right-hand side to have the dimensions of length.

This kind of analysis can never guarantee you that you have the right answer, but it can let you know when you are wrong. And even though it doesn't guarantee you're right, when working with the unknown it is very handy to let dimensional arguments be your guide. They give you a *framework* for fitting the unknown into what you already know about the world.

It is said that fortune favors the prepared mind. Nothing could be more true in the history of physics. And dimensional analysis can prepare our minds for the unexpected. In this regard, the ultimate results of simple dimensional analysis are often so powerful that they can seem magical. To demonstrate these ideas graphically, I want to jump to a modern example based on research at the forefront of physics—where the known and unknown mingle together. In this case, dimensional arguments helped lead to an understanding of one of the four known forces in nature: the "strong" interactions that bind "quarks" together to form protons and neutrons, which in turn make up the nuclei of all atoms. The arguments might seem a little elusive at first reading, but don't worry. I present them because they give you the chance to see explicitly how pervasive and powerful dimensional arguments can be in guiding our physical intuition. The flavor of the arguments is probably more important to carry away with you than any of the results.

Physicists who study elementary-particle physics—that area of physics that deals with the ultimate constituents of matter and the nature of the forces among them—have devised a system of units that exploits dimensional analysis about as far as you can take it. In principle, all three dimensional quantities—length, time, and mass—are independent, but in practice nature gives us fundamental relations among them. For example, if there existed some universal constant that related length and time, then I could express any length in terms of a time by multiplying it by this constant. In fact, nature has been kind enough to provide us with such a constant, as Einstein first showed. The basis of his theory of relativity, which I will discuss later, is the principle that the speed of light, labeled c, is a universal constant, which all observers will measure to have the same value. Since velocity has the dimensions of length/time, if I multiply any time by c, I will arrive at something with the dimension of length—namely, the distance light would travel in this time. It is then possible to express all lengths unambiguously in terms of how long it takes light to travel from one point to another. For example, the distance from your shoulder to your elbow could be expressed as 10^{-9} seconds, since this is approximately the time it takes a light ray to travel this distance. Any observer who measures how far light travels in this time will measure the same distance.

The existence of a universal constant, the speed of light, provides a one-to-one correspondence between any length and time. This allows us to eliminate one of these dimensional quantities in favor of the other. Namely, we can choose if we wish to express all lengths as equivalent times or vice versa. If we want to do this, it is simplest to invent a system of units where the speed of light is numerically equal to unity. Call the unit of length a "light-second" instead of a centimeter or an inch, for example. In this case, the speed of light becomes equal to 1 light-second/second. Now all lengths and their equivalent times will be numerically equal!

We can go one step further. If the numerical values of light-lengths and light-times are equal in this system of units, why consider length and time as being separate dimensional quantities? We could choose instead to equate the dimensions of length and time. In this case all velocities, which previously had the dimensions of length/time, would now be dimensionless, since the dimensions of length and time in the numerator and denominator would cancel. Physically this is equivalent to writing all velocities as a (dimensionless) fraction of the speed of light, so that if I said that something had a velocity of [1/2], this would mean that its velocity was [1/2] the speed of light. Clearly, this kind of system requires the speed of light to be a universal constant for all observers, so that we can use it as a reference value.

Now we have only two independent dimensional quantities, time and mass (or, equivalently, length and mass). One of the consequences of this unusual system is that it allows us to equate other dimensional quantities besides length and time. For example, Einstein's famous formula $E = mc^2$ equates the mass of an object to an equivalent amount of energy. In our new system of units, however, c (= 1) is dimensionless, so that we find that the "dimensions" of energy and mass are now equal. This carries out in practice what Einstein's formula does formally: It makes a one-to-one connection between mass and energy. Einstein's formula tells us that since mass can be turned into energy, we can refer to the mass of something in either the units it had before it was transformed into energy or the units of the equivalent amount of energy that it transforms into. We need no longer speak of the mass of an object in kilograms, or tons, or pounds, but can speak of it in the equivalent units of energy, in, say, "Volts" or "Calories." This is exactly what elementary-particle physicists do when they refer to the mass of the electron as 0.5 million electron Volts (an electron Volt is the energy an electron in a wire gets when powered by a 1-Volt battery) instead of

10^{-31} grams. Since particle-physics experiments deal regularly with processes in which the rest mass of particles is converted into energy, it is ultimately sensible to use energy units to keep track of mass. And that is one of the guidelines: Always use the units that make the most physical sense. Similarly, particles in large accelerators travel at close to the speed of light, so that setting $c = 1$ is numerically practical. This would *not* be practical, however, for describing motions on a more familiar scale, where we would have to describe velocities by very small numbers. For example, the speed of a jet airplane in these units would be about 0.000001, or 10^{-6}.

Things don't stop here. There is another universal constant in nature, labeled *h* and called Planck's constant, after the German physicist Max Planck (one of the fathers of quantum mechanics). It relates quantities with the dimensions of mass (or energy) to those with the dimensions of length (or time). Continuing as before, we can invent a system of units where not only $c = 1$ but $h = 1$. In this case, the relation between dimensions is only slightly more complicated: One finds that the dimension of mass (or energy) becomes equivalent to 1/length, or 1/time. (Specifically, the energy quantity 1 electron Volt becomes equivalent to $1/6 \times 10^{-16}$ seconds.) The net result of all this is that we can reduce the three *a priori* independent dimensional quantities in nature to a single quantity. We can then describe all measurements in the physical world in terms of just one-dimensional quantity, which we can choose to be mass, time, or length at our convenience. To convert between them, we just keep track of the conversion factors that took us from our normal system of units—in which, for example, the speed of light $c = 3 \times 10^8$ meters/sec—to the system in which $c = 1$. For example, volume, with the dimensions of length × length × length = length^3 in our normal system of units, equivalently has the dimensions of 1/mass^3 (or 1/energy^3) in this new system. Making the appropriate conversions to these new units one finds, for example, that a volume of 1 cubic meter (1 meter^3) is equivalent to (1/[10^{-20} electron Volts^3]).

While this is an unusual and new way of thinking, the beauty of it is that with only one fundamental independent dimensional parameter left, we can approximate the results of what may be intrinsically very complicated phenomena simply in terms of a single quantity. In so doing we can perform some magic. For example, say a new elementary particle is discovered that has three times the mass of the proton or, in energy units, about 3 billion electron Volts—3 GeV (Giga electron Volts), for short. If this particle is unstable, what might we expect its lifetime to be before it decays? It may seem impossible to make such an estimate without knowing any of the detailed physical processes involved. However, we can use dimensional analysis to make a guess. The only dimensional quantity in the problem is the rest mass, or equivalently rest energy of the particle. Since the dimensions of time are equivalent to the dimensions of 1/mass in our system, a reasonable estimate of the lifetime would be k/(3 Ge V), where k is some dimensionless number that, in the absence of any other information, we might hope is not too different from 1. We can convert back to our normal units, say, seconds, using our conversion formula (1/1 eV) = 6×10^{-16} sec. Thus we estimate the lifetime of our new particle to be about $k \times 10^{-25}$ seconds.

Of course, there really is no magic here. We have not gotten something for nothing. What dimensional analysis has given us is the *scale* of the problem. It tells us that the "natural lifetime" of unstable particles with this kind of mass is around $k \times 10^{-25}$ seconds, just like the "natural" lifetime of human beings is of the order of k × 75 years. All the real physics (or, in the latter case, biology) is contained in the unknown quantity *k*. If it is very small, or very large, there must be something interesting to be learned in order to understand why.

Dimensional analysis has, as a result, told us something very important. If the quantity *k* differs greatly from 1, we know that the processes involved must be either very strong or very weak, to make the lifetime of such a particle deviate from its natural value as given by dimensional arguments. It would be like seeing a supercow 10 times the size of a normal cow but weighing only 10 ounces. Simple scaling arguments in that case would tell us that such a cow was made of some very exotic material. In fact, many of the most interesting results in physics are those in which naive dimensional scaling arguments break down. What is important to realize is that without these scaling arguments, we might have no idea that anything interesting was happening in the first place!

In 1974, a remarkable and dramatic event took place along these lines. During the 1950s and 1960s, with the development of new techniques to accelerate high-energy beams of particles to collide, first with fixed targets and then with other beams of particles of ever higher energy, a slew of new elementary particles was discovered. As hundreds and hundreds of new particles were found, it seemed as if any hope of simple order in this system had vanished—until the development of the "quark" model in the early 1960s, largely by Murray Gell-Mann at Caltech, brought order out of chaos. All of the new particles that had been observed could be formed out of relatively simple combinations—fundamental objects that Gell-Mann called *quarks.* The particles that were created at accelerators could be categorized simply if they were formed from either three quarks or a single quark and its antiparticle. New combinations of the same set of quarks that make up the proton and neutron were predicted to result in unstable particles, comparable in mass to the proton. These were observed, and their lifetimes turned out to be fairly close to our dimensional estimate (that is, approximately 10^{-25} sec). Generically, the lifetimes of these particles were in the neighborhood of 10^{-24} seconds, so that the constant k in a dimensional estimate would be about 10, not too far from unity. Yet the interactions between quarks, which allow these particles to decay, at the same time seem to hold them so tightly bound inside particles like protons and neutrons that no single free quark had ever been observed. Such interactions seemed so strong as to defy attempts to model them in detail through any calculational scheme.

In 1973, an important theoretical discovery prepared the path. Working with theories modeled after the theory of electromagnetism and the newly established theory of the weak interactions,

David Gross and Frank Wilczek at Princeton and, independently, David Politzer at Harvard discovered that an attractive candidate theory for the strong interactions between quarks had a unique and unusual property. In this theory, each quark could come in one of three different varieties, whimsically labeled "colors," so the theory was called quantum chromodynamics, or QCD. What Gross, Wilczek, and Politzer discovered was that as quarks moved closer and closer together, the interactions between them, based on their "color," should become *weaker and weaker!* Moreover, they proved that such a property was unique to this kind of theory—no other type of theory in nature could behave similarly.

This finally offered the hope that one might be able to perform calculations to compare the predictions of the theory with observations. For if one could find a situation where the interactions were sufficiently weak, one could perform simple successive approximations, starting with noninteracting quarks and then adding a small interaction, to make reliable approximate estimates of what their behavior should be.

While theoretical physicists were beginning to assimilate the implications of this remarkable property, dubbed "asymptotic freedom," experimentalists at two new facilities in the United States—one in New York and one in California—were busily examining ever higher energy collisions between elementary particles. In November 1974, within weeks of each other, two different groups discovered a new particle with a mass three times that of the proton. What made this particle so noticeable was that it had a lifetime about 100 times longer than particles with only somewhat smaller masses. One physicist involved commented that it was like stumbling onto a new tribe of people in the jungle, each of whom was 10,000 years old!

It was soon realized that this new heavy particle had to be made up of a new type of heavy quark (bound to its antiparticle), dubbed the charmed quark, whose existence had in fact been predicted several years earlier by theorists for unrelated reasons. Moreover, the fact that this bound state of quarks lived much longer than it seemed to have any right to could be explained as a direct consequence of asymptotic freedom in QCD. If the heavy quark and antiquark coexisted very closely together in this bound state, their interactions would be weaker than the corresponding interactions of lighter quarks inside particles such as the proton. The weakness of these interactions would imply that it would take longer for the quark and its antiquark to "find" each other and annihilate. Rough estimates of the time it would take, based on scaling the strength of the QCD interaction from the proton size to the estimated size of this new particle, led to reasonable agreement with the observations. QCD had received its first direct confirmation.

In the years since this discovery, experiments performed at still higher energies, where it turns out that the approximations one uses in calculations are more reliable, have confirmed beautifully and repeatedly the predictions of QCD and asymptotic freedom. Even though no one has yet been able to perform a complete calculation in the regime where QCD gets strong, the experimental evidence in the high-energy regime is so overwhelming that no one doubts that we now have the correct theory of the interactions between quarks. And without some dimensional guide for our thinking, the key discoveries that helped put the theory on a firm empirical foundation would not have been appreciated at all. This generalizes well beyond the story of the discovery of QCD. Dimensional analysis provides a framework against which we can test our picture of reality.

If our worldview begins with the numbers we use to describe nature, it doesn't stop there. Physicists insist on also using mathematical relations between these quantities to describe physical processes—a practice that may make you question why we don't use a more accessible language. But we have no choice. Even Galileo appreciated this fact, some 400 years ago, when he wrote: "Philosophy is written in this grand book, the universe, which stands continually open to our gaze. But the book cannot be understood unless one first learns to comprehend the language and read the letters in which it is composed. It is written in the language of mathematics, and its characters are triangles, circles, and other geometric figures without which it is humanly impossible to understand a single word of it; without these, one wanders about in a dark labyrinth."[3]

Now saying that mathematics is the "language" of physics may appear as trite as saying that French is the "language" of love. It still doesn't explain why we cannot translate mathematics as well as we might translate the poems of Baudelaire. And in matters of love, while those of us whose mother tongue is not French may labor under a disadvantage, most of us manage to make do when it counts! No, there is more to it than language alone. To begin to describe how much more, I will borrow an argument from Richard Feynman. Besides being a charismatic personality, Feynman was among the greatest theoretical physics minds in this century. He had a rare gift for explanation, I think due in part to the fact that he had his own way of understanding and deriving almost all the classical results in physics, and also in part to his New York accent.

When Feynman tried to explain the necessity of mathematics,[4] he turned to none other than Newton for a precedent. Newton's greatest discovery, of course, was the Universal Law of Gravity. By showing that the same force that binds us to this sphere we call Earth is responsible for the motions of all the heavenly objects, Newton made physics a universal science. He showed that we have the potential to understand not merely the mechanics of our human condition and our place in the universe but the universe itself. We tend to take it for granted, but surely one of the most remarkable things about the universe is that the same force that guides a baseball out of the park governs the sublimely elegant motion of our Earth around the sun, our sun around the galaxy, our galaxy around its neighbors, and the whole bunch as the universe itself evolves. It didn't have to be that way (or perhaps it did—that issue is still open).

Now Newton's law can be stated in words as follows: The attractive force that gravity exerts between two objects is directed along a line joining them, and depends on the product of their masses and inversely as the square of the distance between them. The verbal explanation is already somewhat cumbersome, but no matter. Combining this with Newton's other law—that bodies react to forces by changing their velocity in the direction of the force, in a manner proportional to the force and inversely proportional to their masses—you have it all. Every consequence of gravity follows from this result. But how? I could give this description to the world's foremost linguist and ask him or her to deduce from this the age of the universe by semantic arguments, but it would probably take longer than this time to get an answer.

The point is that mathematics is also a system of *connections,* created by the tools of logic. For example, to continue with this famous example, Johannes Kepler made history in the early seventeenth century by discovering after a lifetime of data analysis that the planets move around the sun in a special way. If one draws a line between the planet and the sun, then the area swept out by this line as the planet moves in its orbit is always the same in any fixed time interval. This is equivalent (using mathematics!) to saying that when the planet is closer to this sun in its orbit it moves faster, and when it is farther it moves more slowly. But Newton showed that this result is also mathematically identical to the statement that there must be a force directed along a line from the planet to the sun! This was the beginning of the Law of Gravity.

Try as you might, you will never be able to prove, on linguistic grounds alone, that these two statements are identical. But with mathematics, in this case simple geometry, you can prove it to yourself quite directly. (Read Newton's *Principia* or, for an easier translation, read Feynman.)

The point of bringing all this up is not just that Newton might never have been able to derive his Law of Gravity if he hadn't been able to make the mathematical connection between Kepler's observation and the fact that the sun exerted a force on the planets—although this alone was of crucial importance for the advancement of science. Nor is it the fact that without appreciating the mathematical basis of physics, one cannot derive other important connections. The real point is that the connections induced by mathematics are completely fundamental to determining our whole picture of reality.

I think a literary analogy is in order. When I wrote this chapter, I had been reading a novel by the Canadian author Robertson Davies. In a few sentences, he summarized something that hit very close to home: "What really astonished me was the surprise of the men that I could do such a thing. . . . They could hardly conceive that anybody who read . . . could have another, seemingly completely opposite side to his character. I cannot remember a time when I did not take it as understood that everybody has at least two, if not twenty-two, sides to him."[5]

Let me make it a little more personal. One of the many things my wife has done for me has been to open up new ways of seeing the world. We come from vastly different backgrounds. She hails from a small town, and I come from a big city. Now, people who grow up in a big city as I did tend to view other people very differently than people who grow up in a small town. The vast majority of people you meet each day in a big city are one-dimensional. You see the butcher as a butcher, the mailman as a mailman, the doctor as a doctor, and so on. But in a small town, you cannot help but meet people in more than one guise. They are your neighbors. The doctor may be a drunk, and the womanizer next door may be the inspirational English teacher in the local high school. I have come to learn, as did the protagonist in Davies's novel (from a small town!), that people cannot be easily categorized on the basis of a single trait or activity. Only when one realizes this does it become possible truly to understand the human condition.

So, too, every physical process in the universe is multidimensional. It is only by realizing that we can understand each in a host of equivalent, but seemingly different, ways that we can appreciate most deeply the way the universe works. We cannot claim to understand nature when we see only one side of it. And for better or worse, it is a fact that only mathematical relations allow us to see the whole amid the parts. It is mathematics that allows us to say that the world *is* spherical cows.

In one sense, then, mathematics does make the world more complex, by presenting for us the many different faces of reality. But in so doing, it really simplifies our understanding. We need not keep all the faces in our heads at the same time. With the aid of mathematics, we can go from one to the other at will. And if, as I will claim, it is the interconnectedness of physics that ultimately makes it most accessible, then mathematics makes physics accessible.

Moreover, the fact that mathematics allows us to rephrase the same phenomenon in many different guises offers us the continued excitement of discovery. New visions of the same thing are always possible! Also, each new face of reality offers us the possibility of extending our understanding beyond the phenomena that may have led to our new insight. I would be remiss if I did not describe one well-known example of this, which I still find absolutely fascinating twenty-five years after I first learned it from Feynman.

This example involves a familiar but puzzling phenomenon: a mirage. Anyone who has ever driven down a long, straight stretch of highway on a hot summer day will have had the experience of looking down the road and seeing it turn blue in the distance, as if it were wet and reflecting the sky above. This is the less exotic version of the same thing that happens to those poor souls wandering the desert looking for water and seeing it, only to have it disappear as they run toward their vision of salvation.

There is a simple, and standard, explanation of mirages that has to do with the commonly known fact that light bends when it traverses the boundary between two different media. This is why when you are standing in the water, you look shorter than you actually are. The light rays bend at the surface and trick you into thinking your feet are higher up than they are:

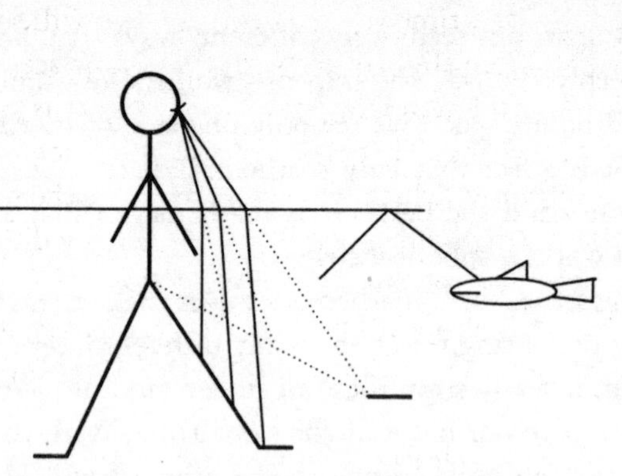

When light goes from a more dense to a less dense medium, as in the picture (going from your legs in the water to your eyes in the air), it always bends "outward." Eventually, if it hits the surface at a big enough angle, it will bend so far that it reflects back into the water. Thus, the shark about to attack remains hidden from sight.

On a still, sultry day, the air right above a road surface gets very hot—much hotter than the temperature of the air higher up. What happens is that the air forms layers, with the most hot *and* the least dense at the bottom, progressing upward to ever cooler and more dense layers. When the light coming from the sky heads toward the road, it gets bent at each layer, until, if there are enough layers, it gets reflected completely until you see it as you look down the road from the car. Thus, the road appears to reflect the blue sky. If you

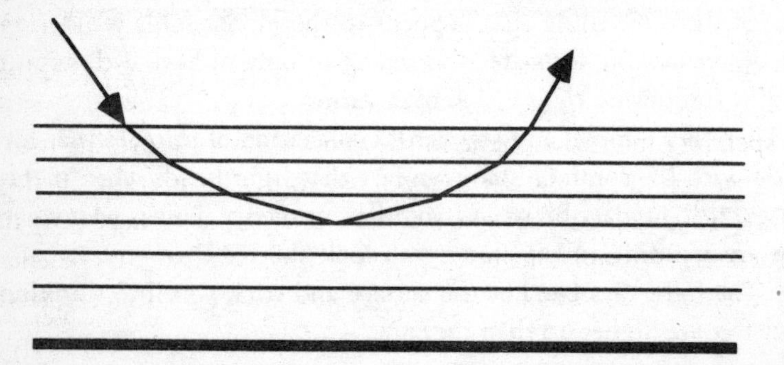

look carefully next time you see a mirage, you will see that the layer of blue comes from slightly above the road's surface.

Now this is the standard explanation, and it is satisfying, if not necessarily inspiring. There is another explanation of the same phenomenon, however, which we now know is mathematically equivalent to this one but which presents a remarkably different picture of how the light gets to your eyes from the sky. It is based on the *principle of least time,* proposed by the French mathematician Pierre de Fermat in 1650, which states that light will always take the path that requires the least time to go from A to B.

This principle clearly is appropriate for the normal travel of light, which is in a straight line. How can it explain mirages, however? Well, light travels faster in a less dense medium (it travels fastest in empty space). Since the air near the road is hotter and less dense, the longer the light stays near the road, the faster it travels. Thus, imagine that a light ray wants to go from point A, to your eye, B. Which path will it take?

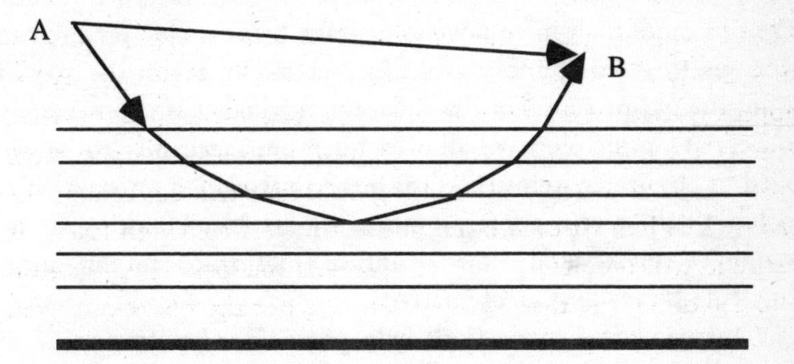

One way to do it would be to travel straight to your eye. In this case, however, the light, while traveling the smallest distance, will be spending most of its time in the dense air high above the road. Another way is to take the path shown in the illustration. In this case, the light travels a longer distance, but it spends more time in the less dense layers near the road, where it travels faster. By balancing the distance traveled with the speed it travels with, you will find that the actual path it takes, the one that produces a mirage, is the one that minimizes the time.

This is strange, if you think about it. How can light determine in advance when it is emitted, which path is the fastest? Does it "sniff" around all possible paths before finally choosing the right one? Clearly not. It just obeys the local laws of physics, which tell it what to do at each interface, and it just *happens,* mathematically, that this turns out always to be the path that takes the shortest time. There is something very satisfying about this finding. It seems more fundamental than the alternative description in terms of the bending of light at different layers in the atmosphere. And in some sense it is. We now understand that the laws of motion of all objects can be re-expressed in a form similar to Fermat's principle for light. Moreover, this new form of expressing the classical Newtonian laws of motion led to a new method, developed by Feynman, for interpreting the laws of quantum mechanics.

By providing different but equivalent ways of picturing the world, mathematics leads to new ways of understanding nature. There is more than mere novelty at stake here. A new picture can allow us to avoid stumbling blocks that might get in the way of using the old picture. For example, the methods based on analogy to Fermat's principle have allowed quantum mechanics to be applied to physical systems that had hitherto remained impenetrable, including the recent effort pioneered by Stephen Hawking to attempt to understand how quantum mechanics might affect Einstein's theory of general relativity.

If mathematical connections help govern our understanding of nature by exposing new ways of picturing the world, this inevitably leads to the following issue I want to leave you with in this chapter. If our abstractions of nature are mathematical, in what sense can we be said to understand the universe? For example, in what sense does Newton's Law explain *why* things move? To turn to Feynman again:

> What do we mean by "understanding" something? Imagine that the world is something like a great chess game being played by the

> gods, and we are observers of the game. We do not know what the rules of the game are; all we are allowed to do is to watch the playing. Of course, if we watch long enough, we may eventually catch on to a few of the rules. The rules of the game are what we call fundamental physics. Even if we knew every rule, however, we might not be able to understand why a particular move is made in the game, merely because it is too complicated and our minds are limited. If you play chess you must know that it is easy to learn all the rules, and yet it is often very hard to select the best move or to understand why a player moves as he does. So it is in nature, only much more so. . . . We must limit ourselves to the more basic question of the rules of the game. If we know the rules, we consider that we "understand" the world.[6]

In the end, we may never go any further than explaining the rules, and may never know why they are as they are. But we have been wonderfully successful at discovering these rules, by abstracting out from complicated situations, where the rules are impossible to trace, to simpler ones, where the rules are self-evident—using the guidance of the tools I have described in this chapter and the last. And when we attempt to understand the world, as physicists, that's all we can hope to do. Nevertheless, if we try very hard and have luck on our side, we can at least have the pleasure of predicting how nature will respond in a situation that has never before been seen. In so doing, we can hope to observe directly the hidden connections in physics that mathematics may first expose and that, in turn, make the world so fascinating.

II

Progress

3

Creative Plagiarism

Plus ça change, plus c'est la même chose.

POPULAR WISDOM might have you believe that new discoveries in science always center on radically new ideas. In fact, most often the opposite is true. The old ideas not only survive but almost always remain seminal. While the universe is infinitely diverse in phenomena, it seems to be rather limited in principles. As a result, in physics there isn't as much premium on new ideas as there is on ideas that work. Thus, one sees the same concepts, the same formalism, the same techniques, the same *pictures,* being twisted and molded and bent as far as possible to apply to a host of new situations, as long as they have worked before.

This might seem to be a pretty timid, even uncreative approach to unlocking nature's secrets, but it isn't. If it takes great boldness to suppose that a slingshot might kill a giant, it is equally bold to suppose that the same methods that determine how far the shot will go might also determine the fate of the universe. It often requires great creativity, too, to see how existing ideas might apply to new and unusual situations. In physics, less is more.

Transplanting old ideas has been successful so regularly that we have come to expect it to work. Even those rare new concepts that have worked their way into physics have been *forced* into existence by the framework of existing knowledge. It is this creative plagiarism that makes physics comprehensible, because it means that the fundamental ideas are limited in number.

Perhaps the greatest modern misconception about science is that scientific "revolutions" do away with all that has gone before. Thus, one might imagine that physics before Einstein is no longer correct. Not so. From here to eternity, the motion of a ball dropped from my hand will be described by Newton's Laws. And while it is the stuff science fiction stories are made of, no new law of physics will ever make the ball fall up! One of the most satisfying aspects of physics is that new discoveries must agree with what is already known to be correct. So, too, the theories of the future will always continue to borrow heavily from those of the past.

This method of doing business complements the notion of approximating reality I discussed earlier. The same "Damn the torpedoes, full speed ahead" mentality suggests that one does not have to understand absolutely everything before moving on. We can probe unknown waters with the tools at our disposal without taking time out to build a new arsenal.

The precedent for this tradition was also created by Galileo. I spoke in the first chapter of Galileo's discovery that concentrating on the simplest aspect of motion and throwing out the irrelevant facts led to a profound reorganization of our picture of reality. What I didn't explain was that he stated right up front that he didn't care *why* things move; all he wanted to do, in his inimitably humble fashion, was investigate *how.* "My purpose is to set forth a very new science dealing with a very ancient subject. There is, in nature, perhaps nothing older than motion, concerning which the books written by philosophers are neither few nor small; nevertheless I have discovered by experiment some properties of it which are worth knowing."[7]

The *how* alone would offer remarkable new insights. As soon as Galileo argued that a body at rest was just a special case of a body moving at a constant velocity, cracks began to emerge in Aristotelian philosophy, which asserted the special status of the former. In fact, Galileo's argument implied that the laws of physics might look the same from the vantage point of an observer moving with constant velocity as they would from the vantage point of one at rest. After all, a third object in constant relative motion with respect to one will also be in constant motion relative to the other. Similarly, an object that speeds up or slows down relative to one will do the same relative to the other. This equivalence between the two vantage points was Galileo's statement of relativity, predating Einstein's by almost three centuries. It is very fortunate for us that it holds, because while we are accustomed to measuring motion as compared to the fixed and stable *terra firma,* all the while the Earth is moving around the sun, and the sun is moving around the galaxy, and our galaxy is moving around in a cluster of galaxies, and so on. So we really are not standing still, but rather moving at some large velocity relative to faraway galaxies. If this background motion had to be taken into account before we could properly describe the physics of a ball flying in the air relative to us on Earth, Galileo and Newton would never have been able to derive these laws in the first place. Indeed, it is only because the constant (on human time scales) motion of our galaxy relative to its neighbors does not alter the behavior of objects moving on Earth that the laws of motion were uncovered, which in turn allowed the developments in astronomy that led to the recognition that our galaxy is moving relative to distant galaxies in the first place.

I'll come back to relativity later. First, I want to describe how Galileo proceeded to milk his first success with uniform motion. Since most motion we see in nature is not in fact uniform, if Galileo was truly to claim to discuss reality, he had to address this issue. Again, he followed his first maxims: Throw out the irrelevant, and don't ask why:

> The present does not seem to be the proper time to investigate the cause of the acceleration of natural motion concerning which various opinions have been expressed by various philosophers, some explaining it by attraction to the center, others to repulsion between the very small parts of the body, while still others attribute it to a certain stress in the surrounding medium which closes in behind the falling body and drives it from one of its positions to another. Now, all these fantasies, and others too, ought to be examined; but it is not really worthwhile. At present it is the purpose of our Author merely to investigate and to demonstrate some of the properties of accelerated motion—meaning thereby a motion, such that . . . its velocity goes on increasing after departure from rest, in simple proportionality to the time, which is the same as saying that in equal time-intervals the body receives equal increments of velocity.[8]

Galileo *defined* accelerated motion to be the simplest kind of nonuniform motion, namely, that in which the velocity of an object changes, but at a constant rate. Is such an idealization relevant? Galileo ingeniously showed that such a simplification in fact described the motion of all falling bodies, if one ignores the extraneous effects of things like air resistance. This discovery paved the way for Newton's Law of Gravity. If there hadn't been a knowledge of the regularity in the underlying motion of falling bodies, the simplicity of assigning a force proportional to the mass of such objects would have been impossible. In fact, to get this far, Galileo had to overcome two other obstacles which are somewhat irrelevant to the point I am making, but his arguments were so simple and clever that I can't resist describing them.

Aristotle had claimed that falling objects instantly acquire their final velocity upon being released. This was a reasonable claim based upon intuitive notions of what we see. Galileo was the first to show convincingly that this was not the case, using a ridiculously simple example. It was based on a *gedanken* or "thought" experiment, in the words of Einstein, a slightly updated version of which I will relate here. Imagine dropping a shoe into a bathtub from 6 inches above the water. Then drop it from 3 feet (and stand back).

If you make the simple assumption that the size of the splash is related to the speed of the shoe when it hits the water, you can quickly convince yourself that the shoe speeds up as it falls.

Next was Galileo's demonstration that all objects fall at the same rate, independent of their mass, if you ignore the effects of air resistance. While most people think of this in terms of the famous experiment of dropping two different objects off the leaning Tower of Pisa, which may actually never have been performed, Galileo in fact suggested a much simpler thought experiment that pointed out the paradox in assuming that objects that are twice as massive fall twice as fast. Imagine dropping two cannonballs of exactly the same mass off a tower. They should fall at the same rate even if their rate of falling did depend upon their mass. Now, as they are falling, imagine that a very skilled and fast craftsman reaching out a window connects them with a piece of strong tape. Now you have a single object whose mass is twice the mass of either cannonball. Common sense tells us that this new object will not suddenly begin falling twice as fast as the two cannonballs were falling before the tape was added. Thus, the rate at which objects fall is *not* proportional to their mass.

Having forced aside these red herrings, Galileo was now ready actually to measure the acceleration of a falling body and show that it is constant. Recall that this implies that the velocity changes at a constant rate. I remind you that in laying the foundation upon which the theory of gravity was developed, Galileo did no more than attempt to describe how things fell, not why. It is like trying to learn about Feynman's game of chess by first carefully examining the configuration of the chessboard and then carefully describing the motion of the pieces. Over and over again since Galileo, we have found that the proper description of the "playing field" on which physical phenomena occur goes a long way toward leading to an explanation of the "rules" behind the phenomena. In the ultimate version of this, the playing field *determines* the rules, and I shall argue later that this is exactly where the thrust of modern physics research is heading . . . but I digress.

Galileo did not stop here. He went on to solve one other major complication of motion by copying what he had already done. To this point he, and we, have discussed motion in only one dimension—either falling down or moving horizontally along. If I throw a baseball, however, it does both. The trajectory of a baseball, again ignoring air resistance, is a curve that mathematicians call a parabola, an arclike shape. Galileo proved this by doing the simplest possible extension of his previous analyses. He suggested that two-dimensional motion could be reduced to two independent copies of one-dimensional motion, which of course he had already described. Namely, the downward component of the motion of a ball would be described by the constant acceleration he had outlined, while the horizontal component of the motion would be described by the constant uniform velocity he had asserted all objects would naturally maintain in the absence of any external force. Put the two together, and you get a parabola.

While this may sound trivial, it both clarified a whole slew of phenomena that are often otherwise misconstrued and set a precedent that has been followed by physicists ever since. First, consider an Olympic long jump, or perhaps a Michael Jordan dunk beginning from the foul-shot line. As we watch these remarkable feats, it is clear to us that the athletes are in the air for what seems an eternity. Considering their speed leading up to the jump, how much longer can they glide in the air? Galileo's arguments give a surprising answer. He showed that horizontal and vertical motion are independent. Thus, if the long-jumper Carl Lewis or the basketball star Michael Jordan were to jump up while standing still, as long as he achieved the same vertical height he achieved at the midpoint of his running jump, he would stay in the air *exactly* as long. Similarly, to use an example preached in physics classes around the world, a bullet shot horizontally from a gun will hit the ground at the same time as a penny dropped while the trigger is being pulled, even if the bullet travels a mile before doing so. The bullet only *appears* to fall more slowly because it moves away so quickly that in the time it takes to leave our sight, neither it nor the penny has had much time to fall at all!

Galileo's success in showing that two dimensions could be just thought of as two copies of one dimension, as far as motion is concerned, has been usurped by physicists ever since. Most of modern physics comes down to showing that new problems can be reduced, by some technique or other, to problems that have been solved before. This is because the list of the types of problems we can solve exactly can probably be counted on the fingers of two hands (and maybe a few toes). For example, while we happen to live in three spatial dimensions, it is essentially impossible to solve exactly most fully three-dimensional problems, even using the computational power of the fastest computers. Those we can solve invariably involve either effectively reducing them to solvable one- or two-dimensional problems by showing that some aspects of the problem are redundant, or at the very least reducing them to independent *sets* of solvable one- or two-dimensional problems by showing that different parts of the problem can be treated independently.

Examples of this procedure are everywhere. I have already discussed our picture of the sun, in which we assume that the internal structure is the same throughout the entire sun at any fixed distance from the center. This allows us to turn the interior of the sun from a three-dimensional problem to an effectively one-dimensional problem, described completely in terms of the distance, r, from the solar center. A modern example of a situation in which we don't ignore but rather break up a three-dimensional problem into smaller pieces can be obtained closer to home. The laws of quantum mechanics, which govern the behavior of atoms and the particles that form them, have allowed us to elucidate the laws of chemistry by explaining the structure of atoms, which make up all materials. The simplest atom is the hydrogen atom, with only a single, positively charged particle at its center, the proton, surrounded by a single negatively charged particle, an electron. The quantum-mechanical solution of the behavior of even such a simple system is quite rich. The electron can exist in a set of discrete states of differing total energy. Each of the main "energy levels" is itself subdivided into states in which the shape of the electron's "orbits" are different. All of the intricate behavior of chemistry—responsible for the biology of life, among other things—reflects, at some basic level, the simple counting rules for the number of such available states. Elements with all of these states but one in a certain level filled by electrons like to bond chemically to elements that have only one lone electron occupying the highest energy levels. Salt, also called sodium chloride, for example, exists because sodium shares its lone electron with chlorine, which uses it to fill up the otherwise sole unoccupied state in its highest energy level.

The only reason we have been able to enumerate the level structure of even the simplest atoms such as hydrogen is because we have found that the three-dimensional nature of these systems "separates" into two separate parts. One part involves a one-dimensional problem, which relates to understanding simply the radial distance of the electron from the proton. The other part involves a two-dimensional problem, which governs the angular distribution of the electron "orbits" in the atom. Both of these problems are solved separately and then combined to allow us to classify the total number of states of the hydrogen atom.

Here's a more recent and more exotic example, along similar lines. Stephen Hawking has become known for his demonstration in 1974 that black holes are not black—that is, they emit radiation with a temperature characteristic of the mass of the black hole. The reason this discovery was so surprising is that black holes were so named because the gravitational field at their surface is so strong that nothing inside can escape, not even light. So how can they emit radiation? Hawking showed that, in the presence of the strong gravitational field of the black hole, the laws of quantum mechanics allow this result of classical thinking to be evaded. Such evasions of classical "no go" theorems are common in quantum mechanics. For example, in our classical picture of reality, a man resting in a valley between two mountains might never be able to get into a neighboring valley without climbing over one or the other of the mountains. However, quantum mechanics allows an electron in an atom, say, with an energy smaller than that required to escape from the atom, according to classical principles, sometimes to "tunnel" out from inside the electric field binding it, and find itself finally free of its former chains! A standard example of this phenomenon is radioactive decay. Here, the configuration of particles—protons and neutrons—buried deep in the nucleus of an atom can suddenly change. Depending upon the properties of an individual atom or nucleus, quantum mechanics tells us that it is possible for one or more of these particles to escape from the nucleus, even though classically they are all irretrievably bound there. In another example, if I throw a ball at a window, either the ball will have enough energy go through it, or it will bounce off the window and return. If the ball is small enough so that its behavior is governed by quantum-mechanical principles, however, things are different. Electrons, say, impinging on a thin barrier can do both! In a more familiar example, light impinging on the surface of a material like a mirror might normally be reflected. If the mirror is thin enough, however, we find that even though most is reflected, some of the light can "tunnel" through the mirror and appear on the other side! (I shall outline the new "rules" that govern this weird behavior later. For the moment, take it as a given.)

Hawking showed that similar phenomena can occur near the black hole. Particles can tunnel through the gravitational barrier at the surface of the black hole and escape. This demonstration was a tour de force because it was the first time the laws of quantum mechanics had been utilized in connection with general relativity to reveal a new phenomenon. Again, however, it was possible only because, like the hydrogen atom, the quantum-mechanical states of particles around a black hole are "separable"—that is, the three-dimensional calculation can be effectively turned into a one-dimensional problem and an independent two-dimensional problem. If it weren't for this simplification, we might still be in the dark about black holes.

Interesting as these technical tricks might be, they form only the tip of the iceberg. The real reason we keep repeating ourselves as we discover new laws is not so much due to our character, or lack thereof, as it is due to the character of nature. She keeps repeating herself. It is for this reason that we almost universally check to see whether new physics is really a reinvention of old physics. Newton, in discovering his Universal Law of Gravity, benefited tremendously from the observations and analyses of Galileo, as I have described. He also benefited from another set of careful observations by the Danish astronomer Tycho Brahe, as analyzed by his student Johannes Kepler—a contemporary of Galileo.

Both Brahe and Kepler were remarkable characters. Brahe, from a privileged background, became the most eminent astronomer in Europe after his observations of the supernova of 1572. He was given an entire island by the Danish monarch King Frederick II to use as an observatory site, only to be forced to move some years later by Frederick's successor. Unhindered by, or perhaps because of, his arrogance (and a false nose made of metal), Brahe managed to improve in one decade the precision in astronomical measurement by a factor of 10 over what which it had maintained for the previous thousand years—and all of this without a telescope! In Prague, where he had gone in exile from Denmark, Brahe hired Kepler a year before his own death to perform the intricate calculational analysis required to turn his detailed observations of planetary motions into a consistent cosmology.

Kepler came from another world. A child of a family of modest means, his life was always at the edge, both financially and emotionally. Besides his scientific pursuits, Kepler found time to defend his mother successfully from prosecution as a witch and to write what was probably the first science fiction novel, about a trip to the moon. In spite of these diversions, Kepler approached the task of analyzing the data in Brahe's notebooks, which he inherited upon Brahe's death, with an uncommon zeal. Without so much as a Macintosh, much less a supercomputer, he performed a miracle of complicated data analysis that would occupy the better part of his career. From the endless tables of planetary positions, he arrived at the three celebrated laws of planetary motion that still bear his name, and that provided the key clues Newton would use to unravel the mystery of gravity.

I mentioned one of Kepler's Laws earlier—namely, that the orbits of the planets sweep out equal areas in equal times—and how Newton was able to use this to infer that there was a force pulling the planets towarɑ ..e sun. We are so comfortable with this idea nowadays that it is worth pointing out how counterintuitive it really is. For centuries before Newton, it was assumed that the force needed to keep the planets moving around the sun must emanate from something *pushing* them around. Newton quite simply relied on Galileo's law of uniform motion to see that this was unnecessary. Indeed, he argued that Galileo's result that the motion of objects thrown in the air would trace out a parabola, and that their horizontal velocity would remain constant, would imply that an object thrown sufficiently fast could orbit the Earth. Due to the curvature of the Earth an object could continue to "fall" toward the earth, but if it were moving fast enough initially, its constant horizontal motion could carry it far enough that in "falling" it would continue to remain a constant distance from the Earth's surface. This is demonstrated in the following diagram, copied from Newton's *Principia:*

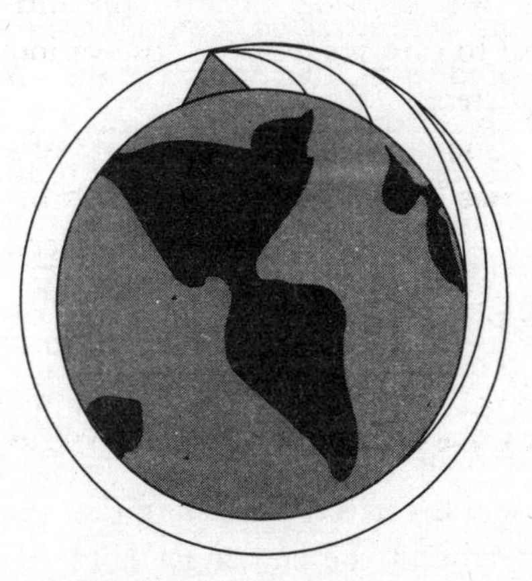

Having recognized that a force that clearly pulled downward at the Earth could result in a body continually falling toward it for eternity—what we call an orbit—it did not require too large a leap of imagination to suppose that objects orbiting the sun, such as the planets, were being continually pulled toward the sun, and not pushed around it. (Incidentally, it is the fact that objects in orbit are continually "falling" that is responsible for the weightlessness experienced by astronauts. It has nothing to do with an absence of gravity, which is almost as strong out at the distances normally traversed by satellites and shuttles as it is here on Earth.)

In any case, another of Kepler's laws of planetary motion provided the icing on the cake. This law yielded a quantitative key that unlocked the nature of the gravitational attraction between objects, for it gave a mathematical relation between the length of each planet's year—the time it takes for it to go around the sun and its distance from the sun. From this law, one could easily derive that the velocity of the planets around the sun falls in a fixed way with their distance from the sun. Specifically, Kepler's laws showed that their velocity falls inversely with the square root of their distance from the sun.

Armed with this knowledge, and his own generalization from the results of Galileo that the acceleration of moving bodies must be proportional to the force exerted on them, Newton was able to show that if planets were attracted toward the sun with a force proportional to the product of their mass and the sun's mass, divided by the square of the distance between them, Kepler's velocity law would naturally result. Moreover, he was able to show that the constant of proportionality would be precisely equal to the mass of the sun times the strength of the gravitational force. If the strength of the gravitational force between all objects is universal, this could be represented by a constant, which we now label G.

Even though it was beyond the measuring abilities of his time to determine the constant G directly, Newton did not need this to prove that his law was correct. Reasoning that the same force that held the planets around the sun must hold the moon in orbit around the Earth, he compared the predicted motion of the moon around the Earth—based on extrapolating the measured downward acceleration of bodies at the earth's surface with the actual measured motion: namely, that it takes about 28 days for the moon to orbit the Earth. The predictions and the observations agreed perfectly. Finally, the fact that the moons of Jupiter, which Galileo had first discovered with his telescope, also obeyed Kepler's law of orbital motion, this time vis-à-vis their orbit around Jupiter, made the universality of Newton's Law difficult to question.

Now, I mention this story not just to reiterate how the mere observation of *how* things move—in this case, the planets—led to an understanding of *why* they move. Rather, it is to show you how we have been able to exploit these results even in modern research. I begin with a wonderful precedent created by the British scientist Henry Cavendish, about 150 years after Newton discovered the Law of Gravity.

When I graduated and became a postdoctoral fellow at Harvard University, I quickly learned a valuable lesson there: Before writing a scientific paper, it is essential to come up with a catchy title. I thought at the time that this was a recent discovery in science, but I have since learned that it has a distinguished tradition, going back at least as far as Cavendish in 1798.

Cavendish is remembered for performing the first experiment that measured directly in the laboratory the gravitational attraction between two known masses, thus allowing him to measure, *for the first time,* the strength of gravity and determine the value of G. In reporting his results before the Royal Society he didn't entitle his paper "On Measuring the Strength of Gravity" or "A Determination of Newton's Constant G." No, he called it "Weighing the Earth."

There was a good reason for this sexy title. By this time Newton's Law of Gravity was universally accepted, and so was the premise that this force of gravity was responsible for the observed motion of the moon around the Earth. By measuring the distance to the moon (which was easily done, even in the seventeenth century, by observing the change in the angle of the moon with respect to the horizon when observed at the same time from two different locations—the same technique surveyors use when measuring distances on Earth), and knowing the period of the moon's orbit—about 28 days—one could easily calculate the moon's velocity around the Earth. Let me reiterate that Newton's great success involved not just his explanation of Kepler's Law that the velocity of objects orbiting the sun was inversely proportional to the square root of their distance from the sun. He also showed that this same law could apply to the motion of the moon and to objects falling at the Earth's surface. His Law of Gravity implied that the constant of proportionality was equal to the product of *G* times the mass of the sun in the former case, and *G* times the mass of the Earth in the latter. (He never actually proved that the value of *G* in the two cases is, in fact, the same. This was a guess, based on the assumption of simplicity and on the experimental observation that the value of *G* seemed to be the same for objects falling at the Earth's surface as for the moon, and that the value of *G* that applied to the planets orbiting the sun appeared uniform for all these planets. Thus, a simple extrapolation suggested that a single value of *G* might suffice for everything.)

In any case, by knowing the distance of the moon from the Earth, and the velocity of the moon around the Earth, you could plug in Newton's Law and determine the *product* of *G* times the mass of the Earth. Until you knew independently the value of *G*, however, you could not extract from this the mass of the Earth. Thus, Cavendish, who was the first person to determine the value of *G*, 150 years after Newton had proposed it, was also the first to be able to determine the mass of the Earth. (The latter sounds a lot more exciting, and so his title.)

We have benefited not only from Cavendish's astute recognition of the value of good press but also from the technique he pioneered for weighing the Earth by pushing Newton's Law as far as he could. It remains in use today. Our best measurement of the mass of the sun comes from exactly the same procedure, using the known distances and orbital velocities of each of the planets. In fact, this procedure is so good that we could, in principle, measure the mass of the sun to one part in a million, based on the existing planetary data. Unfortunately, Newton's constant G is the poorest measured fundamental constant in nature. We know it to an accuracy of only 1 part in 10,000 or so. Thus, our knowledge of the sun's mass is limited to this accuracy.

Nevertheless when you have a good thing going, don't stop. Our sun (and thus our solar system) orbits around the outer edge of the Milky Way galaxy, and so we can use the sun's known distance from the center of the galaxy (about 25,000 light years) and its known orbital velocity (about 150 miles/second), to "weigh" the galaxy. When we do this, we find that the mass of material enclosed by our orbit corresponds to roughly a hundred billion solar masses. This is heartening, since the total light emitted by our galaxy is roughly equivalent to that emitted by about a hundred billion stars more or less like our Sun. (Both of these observations provide the rationale for my previous statement that there are about a hundred billion stars in our galaxy.)

A remarkable thing happens when we try to extend this measurement by observing the velocity of objects located farther and farther from the center of our galaxy. Instead of falling off, as it should if all the mass of our galaxy is concentrated in the region where the observed stars are, this velocity remains constant. This suggests that, instead, there is ever more mass located outside the region where the stars shine. In fact, current estimates suggest that there is at least ten times more stuff out there than meets the eye! Moreover, similar observations of the motions of stars in other galaxies all suggest the same thing. Extending Newton's Law further to use the observed motion of galaxies themselves amid groups and clusters of galaxies confirms this notion. When we use Newton's Law to weigh the universe, we find that at least 90 percent of it is "dark."

Observations that the universe appears to be dominated by what we call *dark matter* are at the heart of one of the most exciting and actively pursued mysteries in modern physics. It would require an entire book to describe adequately the efforts to determine what this stuff might be (and, coincidentally, I have already written one). Here, however, I just want to make you aware of it and to demonstrate that this very modern research problem stems from exploiting exactly the same analysis as Cavendish used over two centuries ago to weigh the Earth for the first time.

At this point, you might be tempted to ask why we believe we can push Newton's Law this far. After all, requiring a whole new source of nonluminous matter to fill the universe seems like a lot to ask for. Why not assume instead that Newton's Law of Gravity doesn't apply on galactic scales and larger? It might seem strange at first, but I hope my arguments to this point will help make it more understandable why physicists might believe that postulating a universe filled with dark matter is more conservative than throwing out Newton's Law. Newtonian gravity has worked perfectly thus far to explain the motion of literally everything under the sun. We have no reason to believe it might not apply to larger scales. Moreover, there is a distinguished tradition of overcoming other possible challenges to Newton's Law. For example, after the planet Uranus was discovered, it was recognized that the motion of this object, the farthest object from the sun in our solar system known at the time, could not be accounted for by Newtonian gravity based on the attraction of the sun and the other planets. Could this be the first hint of breakdown in the Universal Law? Yes, but it was simpler to suppose that its observed motion might be affected by some as yet unseen "dark" object. Careful calculations using Newton's Law performed in the eighteenth century pinpointed where such an object might be. When telescopes were pointed at this region, the planet Neptune was soon discovered. Similar later observations of Neptune's motion led to the discovery of Pluto in 1930.

An even earlier example points out the utility of sticking with a law that works. Often confronting the apparent challenges to such a law can lead to exciting new physical discoveries that are not directly related to the law itself. For example, in the seventeenth century the Danish astronomer Ole Roemer observed the motion of the moons of Jupiter and discovered a curious fact. At a certain time of year, the moons reappear from behind Jupiter about four minutes earlier than one would expect from applying Newton's Law directly. Six months later, the moons are four minutes late. Roemer deduced that this was not a failure of Newton's Law, but rather an indication of the fact that light travels at a finite speed. You may remember that light traverses the distance between the Earth and the sun in about eight minutes. Thus, at one time of year, the Earth is eight "light-minutes" closer to Jupiter than it is when it is on the other side of its orbit around the sun. This accounts for the eight-minute difference in timing the orbits of Jupiter's moons. In this way, Roemer was actually able to estimate the speed of light accurately, over 200 years before it was measured directly.

Thus, while we don't know that we can go on weighing larger and larger regions of the universe as we have already weighed the Earth and the sun, doing so is currently the best bet. It also offers the greatest hope for progress. A single definitive observation may be enough to disprove a theory in physics. But the observed motion of objects in our galaxy and other galaxies is not such a definitive test. It can be explained by the existence of dark matter, for which there is now concurrent support coming from arguments about the formation of large-scale structures in the universe. Further observations will tell us whether our stubborn persistence has been well founded, and in so doing we may discover what most of the universe is made from.

I received numerous letters after writing the book about dark matter from individuals who were convinced that the observations I described provided definitive support for their bold new theories, which they claimed "professionals" have been too narrow-minded to consider. I wish I could convince them that open-mindedness in physics involves a stubborn allegiance to well-proven ideas until there is definitive evidence that they must be transcended. Most of the vital revolutions in this century were not based on discarding old ideas as much as on attempting to accommodate them and using the resultant wisdom to confront existing experimental or theoretical puzzles. In the words of Feynman again, himself one of the most original physicists of our time: "Scientific creativity is imagination in a straitjacket."[9]

Consider perhaps the most famous revolution in physics in this century: Einstein's development of the special theory of relativity. While there is no denying that the result of special relativity was to force a total revision in our notions of space and time, the origin of this idea was a less ambitious attempt to make consistent two well-established physical laws. In fact, the entire thrust of Einstein's analysis was an effort to rework modern physics into a mold in which it would accommodate Galileo's relativity principle, developed some three hundred years earlier. Seen in this way, the logic behind Einstein's theory can be framed quite simply. Galileo argued that the existence of uniform motion required that the laws of physics, as measured by any uniformly moving observer—including one standing still—should be identical. This implies a surprising result: It is impossible to perform any experiment that proves definitively that you are at rest. Any observer moving at a constant velocity with respect to any other observer can claim that he or she is at rest and the other is moving. No experiment that either can perform will distinguish which is moving. We have all had this experience. As you watch the train on the next track from the one you are on depart, it is sometimes hard to tell at first which train is moving. (Of course, if you are traveling on a train in the United States, it quickly becomes easy. Just wait to feel the bumps.)

In perhaps the major development of nineteenth-century physics, James Clerk Maxwell, the preeminent theoretical physicist of his time, put the final touches on a complete theory of electromagnetism, one that explained consistently all of the physical phenomena that now govern our lives—from the origin of electric currents to the laws behind generators and motors. The crowning glory of this theory was that it "predicted" that light must exist, as I shall describe.

The work of other physicists in the early part of the nineteenth century, in particular, the British scientist Michael Faraday—a former bookbinder's apprentice who rose to become director of that centerpiece of British science, the Royal Institution—had established a remarkable connection between electric and magnetic forces. At the beginning of the century, it appeared that these two forces, which were well known to natural philosophers, were distinct. Indeed, on first glimpse, they are. Magnets, for example, always have two "poles," a north and a south. North poles attract south poles and vice versa. If you cut a magnet in half, however, you do not produce an isolated north or south pole. You produce two new smaller magnets, each of which has two poles. Electric charge, on the other hand, comes in two types, named positive and negative by Ben Franklin. Negative charges attract positive ones and vice versa. However, unlike magnets, positive and negative charges can be easily isolated.

Over the course of the first half of the century, new connections between electricity and magnetism began to emerge. First it was established that magnetic fields, that is, magnets, could be created by moving electric charges, that is, currents. Next it was shown that a magnet would deflect the motion of a moving electric charge. In a much bigger surprise, it was shown (by Faraday and, independently, by the American physicist Joseph Henry) that a moving magnet can actually create an electric field and cause a current to flow.

There is an interesting story associated with the latter, which I can't resist relating (especially in these times of political debate over funding such projects as the Superconducting Supercollider, of which I shall speak later). Faraday, as director of the Royal Institution, was performing "pure" research—that is, he was attempting to discover the fundamental nature of electric and magnetic forces, not necessarily in the search for possible technological applications. (This was probably before the era in which such a distinction was significant, in any case.) In fact, however, essentially all of modern technology was made possible because of this research: the principle behind which all electric power is generated today, the principles behind the concept of the electric motor, and so on. During Faraday's tenure as director of the Royal Institution, his laboratory was visited by the Prime Minister of England, who bemoaned this abstract research and wondered aloud whether there was any use at all in these gimmicks being built in the lab. Faraday replied promptly that these results were very important, so important that one day Her Majesty's government would tax them! He was right.

Returning to the point of this history, by the middle of the nineteenth century it was clear that there was some fundamental relationship between electricity and magnetism, but no unified picture of these phenomena was yet available. It was Maxwell's great contribution to unify the electric and magnetic forces into a single theory—to show that these two distinct forces were really just different sides of the same coin. In particular, Maxwell extended the previous results to argue very generally that any changing electric field would create a magnetic field, and, in turn, any changing magnetic field would create an electric field. Thus, for example, if you measure an electric charge at rest, you will measure an electric field. If you run past the same charge, you will also measure a magnetic field. Which you see depends upon your state of motion. One person's electric field is another person's magnetic field. They are really just different aspects of the same thing!

As interesting as this result was for natural philosophy, there was another consequence that was perhaps more significant. If I jiggle an electric charge up and down, I will produce a magnetic field due to the changing motion of the charge. If the motion of the charge is itself continually changing, I will in fact produce a *changing* magnetic field. This changing magnetic field will in turn produce a changing electric field, which will in turn produce a changing magnetic field, and so on. An "electromagnetic" disturbance, or wave, will move outward. This is a remarkable result. More remarkable perhaps was the fact that Maxwell could calculate, based purely on the measured strengths of the electric and magnetic forces between static and moving charges, how fast this disturbance should move. The result? The wave of moving electric and magnetic fields should propagate at a speed identical to that at which light is measured to travel. Not surprisingly, in fact, it turns out that light is nothing other than an electromagnetic wave, whose speed is fixed in terms of two fundamental constants in nature: the strength of the electric force between charged particles and the strength of the magnetic force between magnets.

I cannot overstress how significant a development this was for physics. The nature of light has played a role in all the major developments of physics in this century. For the moment I want to focus on just one. Einstein was, of course, familiar with Maxwell's results on electromagnetism. To his great credit, he also recognized clearly that they implied a fundamental paradox that threatened to overthrow the notion of Galilean relativity.

Galileo told us that the laws of physics should be independent of where one measures them, as long as you are in a state of uniform motion. Thus, for example, two different observers, one in a laboratory on a boat floating at a constant velocity downstream and one in a laboratory fixed on shore, should measure the strength of the electric force between fixed electric charges located 1 meter apart in their respective laboratories to be exactly the same. Similarly, the force between two magnets 1 meter apart should be measured to be the same independent of which lab one performs the measurement in.

On the other hand, Maxwell tells us that if we jiggle a charge up and down, we will always produce an electromagnetic wave that moves away from us at a speed fixed by the laws of electromagnetism. Thus, an observer on the boat who jiggles a charge will see an electromagnetic wave travel away at this speed. Similarly, an observer on the ground who jiggles a charge will produce an electromagnetic wave that moves away from him or her at this speed. The only way these two statements can apparently be consistent is if the observer on the ground measures the electromagnetic wave produced by the observer on the boat to have a different velocity than the wave he or she produces on the ground.

But, as Einstein realized, there was a problem with this. Say I am "riding" next to a light wave, he proposed, at almost the speed of this wave. I imagine myself at rest, and as I look at a spot fixed to be next to me, which has an electromagnetic wave traveling slowly across it, I see a changing electric and magnetic field at that spot. Maxwell tells me that these changing fields should generate an electromagnetic wave traveling outward at the speed fixed by the laws of physics, but instead all I see is a wave moving slowly past me.

Einstein was thus faced with the following apparent problem. Either give up the principle of relativity, which appears to make physics possible by saying that the laws of physics are independent of where you measure them, as long as you are in a state of uniform motion; or give up Maxwell's beautiful theory of electromagnetism and electromagnetic waves. In a truly *revolutionary* move, he chose to give up neither. Instead, knowing that these fundamental ideas were too sensible to be incorrect, he made the bold decision that, instead, one should consider changing the notions of space and time themselves to see whether these two apparently contradictory requirements could be satisfied at the same time.

His solution was remarkably simple. The only way that both Galileo and Maxwell could be correct at the same time would be if both observers measured the speed of the electromagnetic waves they themselves generated to be the value predicted by Maxwell, *and* if they also measured the speed of the waves generated by their counterpart *also* to be this same speed. Thus, this must be what happens!

This one requirement may not sound strange, but think for a moment about what it suggests. If I watch a child in a car moving past me throw up, I will measure the speed of the vomit with respect to me to be the speed of the car, say, 60 miles per hour, plus the speed of the vomit with respect to the car, say, 5 feet per second. The mother in the front seat of the car, however, will measure the speed of the vomit with respect to her to be just the latter, 5 feet per second. However, if instead of vomiting, the child shines a laser beam on his mother, Einstein tells me the result will be different. Special relativity appears to require that I measure the speed of the light ray with respect to me to be the speed Maxwell calculated, *not* this speed plus 60 miles per hour. Similarly, the child's mother will also measure the same speed.

The only way this can be possible is if somehow our measurements of space and time "adjust" themselves so that both of us measure the same speed. After all, speed is measured by determining how far something travels in a fixed time interval. If either the ruler in the car used to measure distance "shrinks" with respect to mine, or the clock that ticks to measure time runs slowly with respect to mine, then it would be possible for both of us to record the same speed for the light ray. In fact, Einstein's theory says that *both* happen! Moreover, it states that things are perfectly reciprocal. Namely, as far as the woman in the car is concerned, my ruler "shrinks" with respect to hers and my clock runs slow!

These statements sound so absurd that no one believes them on first reading. In fact, it requires a far more detailed analysis to investigate fully all the implications of Einstein's claim that the speed of light must be measured to be the same for all observers and to sort out all the apparent paradoxes it implies. Among these implications are the *now measured* facts that moving clocks slow down, that moving particles appear more massive, and that the speed of light is the ultimate speed limit—nothing physical can move faster. These follow logically from the first assertion. While Einstein no doubt deserves credit for having the courage and fortitude to follow up on all these consequences, the really difficult task was coming up with his claim about the constancy of light in the first place. It is a testimony to his boldness and creativity *not* that he chose to throw out existing laws that clearly worked, but rather that he found a creative way to live within their framework. So creative, in fact, that it sounds nuts.

In the next chapter I'll come back to a way of viewing Einstein's theory so that it appears less crazy. For now, however, I want to leave this as a reminder for anyone who has wanted to use the claim that "they said Einstein was crazy too!" to validate his or her own ideas: What Einstein precisely did *not* do was to claim that the proven laws of physics that preceded him were wrong. Rather, he showed that they implied something that hadn't before been appreciated.

The theory of special relativity, along with quantum mechanics, forced revisions in our intuitive picture of reality more profoundly than any other developments in the twentieth century. Between the two of them, they shook the foundations of what we normally consider reasonable by altering how we understand those pillars of our perception: space, time, and matter. To a great extent, the rest of this century has been involved with coming to grips with the implications of these changes. This has sometimes required just as much gumption and adherence to proven physics principles as was required to develop the theories themselves. Here's a case in point related to the subtle marriage of quantum mechanics with special relativity that I alluded to in my discussion of the Shelter Island meeting: the creation of particle-antiparticle pairs from nothing.

While I have mentioned quantum mechanics several times, I have not yet discussed its tenets in any kind of detail, and there is good reason for this. The road to its discovery was much less direct than that for relativity, and also the phenomena to which it applies—at the realm of atomic and subatomic physics—are less familiar. Nevertheless, as the dust settles, we now recognize that quantum mechanics, too, derives from a single, simply stated assertion—which also seems crazy. If I throw a ball in the air and my dog catches it 20 feet away, I can watch the ball during its travels and check that the trajectory it follows is that predicted by Galilean mechanics. However, as the scale of distances and travel times gets smaller, this certainty slowly disappears. The laws of quantum mechanics assert that if an object travels from A to B, you cannot assert that it definitely traverses any particular point in between!

One's natural reaction to this claim is that it is immediately disprovable. I can shine a light on the object and *see* where it goes! If you do shine a "light" between A and B, you can detect the object, say an electron, at some particular point, C, between the two. For example, if a series of electron detectors are set along a line separating A and B, only one of them will click as the particle passes by.

So what happens to the original assertion if I can apparently so easily refute it? Well, nature is subtle. I can surely detect the passage of a particle such as an electron, but I cannot do so with impunity! If, for example, I send a beam of electrons toward a phosphorescent screen, like a TV screen, they will light up areas of the screen as they hit it. I can then put up a barrier on the way to the screen with two nearby narrow slits in the way of the beam, so that the electrons must go through one or the other to make it to the screen. In order to say which one each individual electron goes through, I can set up a detector at each of the slits. The most remarkable thing then happens. If I don't measure the electrons as they pass through the slits, I see one pattern on the screen. If I measure them one by one, so I can ascertain which trajectory each takes, the pattern I see on the screen changes. Doing the measurement changes the result! Thus, while I can confidently assert that each of the electrons I detect does in fact pass through one of the slits, I cannot from this make any inference about the electrons I *don't* detect, which clearly have a different behavior.

Behavior such as this is based on the fact that the laws of quantum mechanics require at a certain fundamental level an intrinsic uncertainty in the measurement of natural processes. For example, there is an absolute limit on our ability to measure the position of a moving particle and at the same time to know its speed (and hence where it is going). The more accurately I measure one, the less accurately I can know the other. The act of measurement, because it disturbs a system, changes it. On normal human scales, such disturbances are so small as to go unnoticed. But on the atomic scale, they can become important. Quantum mechanics gets its name because it is based on the idea that energy cannot be transported in arbitrarily small amounts, but instead comes in multiples of some smallest "packet" or *quanta* (from the German). This smallest packet is comparable to the energies of particles in atomic systems, and so when we attempt to measure such particles, we invariably must do so by allowing some signal to be transferred that is of the same order of magnitude as their initial energy. After the transfer, the energy of the system will be changed, and so will the particle motions involved. If I measure a system over a very long period, the average energy of the system will remain fairly constant, even if it changes abruptly from time to time throughout the measurement process. Thus, one arrives at another famous "uncertainty relation": The more accurately I want to measure the energy of a system, the longer I have to measure it.

These uncertainty relations form the heart of quantum-mechanical behavior. They were first elucidated by the German physicist Werner Heisenberg, one of the founders of the theory of quantum mechanics. Heisenberg, like the other boy wonders involved in the development of this theory during the 1920s and 1930s, was a remarkable physicist. Some of my colleagues insist that he is second only to Einstein in his impact on physics in this century. Unfortunately, however, Heisenberg's popular reputation today is somewhat tainted because he remained a prominent scientific figure even during the days of Nazi Germany. It is not at all clear that he overtly supported the Nazi regime or its war effort. But, unlike a number of his colleagues, he did not actively work against it. In any case, his work on quantum mechanics—in particular, his elucidation of the uncertainty principle—changed forever the way we understand the physical world. In addition, probably no physics result in this century has so strongly affected philosophy.

Newtonian mechanics implied complete determinism. The laws of mechanics imply that one could, in principle, completely predict the future behavior of a system of particles (presumably including the particles that make up the human brain) with sufficient knowledge of the positions and motions of all particles at any one time. The uncertainty relations of quantum mechanics suddenly changed all that. If one took a snapshot giving precise information on the positions of all particles in a system, one would risk losing all information about where those particles were going. With this apparent loss in determinism—no longer could one make completely accurate predictions about the future behavior of all systems, even in *principle*—came, at least in many people's minds, free will.

While the principles of quantum mechanics have excited many nonphysicists, especially philosophers, it is worth noting that all the philosophical implications of quantum mechanics have very little impact whatsoever on physics. All that physicists need to consider are the rules of the game. And the rules are that inherent, and calculable, measurement uncertainties exist in nature. There are many ways of attempting to describe the origin of these uncertainties, but, as usual, the only completely consistent ones (and there are, as usual, a number of different but equivalent ones) are mathematical. There is one mathematical formulation that is particularly amenable to visualization, and it is due to none other than Richard Feynman.

One of Feynman's greatest contributions in physics was to reinterpret the laws of quantum mechanics in terms of what is known in mathematical parlance as *path integrals* along the lines of Fermat's principle for light that I discussed in the last chapter. What started as a "mere" calculational scheme has now influenced the way a whole generation of physicists picture what they are doing. It even introduced a mathematical trick, called "imaginary time," that Stephen Hawking has alluded to in his popular *A Brief History of Time.*

Feynman's path integrals give rules for calculating physical processes in quantum mechanics, and they go something like this. When a particle moves from point A to point B, imagine all the possible paths it can take:

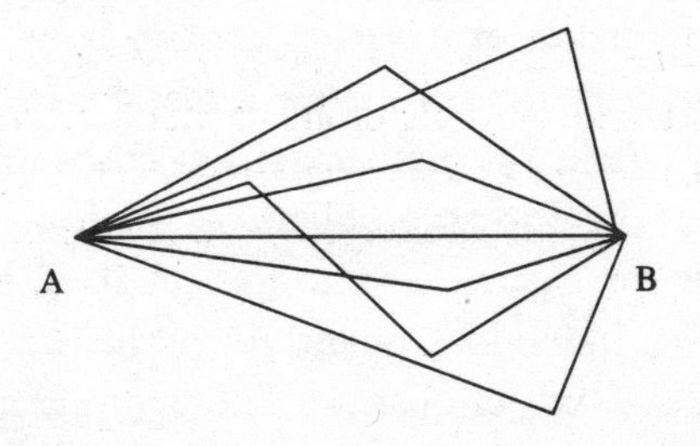

With each path, one associates a kind of probability that the particle will take it. The tricky part is to calculate the probability associated with a given path, and that is what all the mathematical tools such as imaginary time are for. But that is not what concerns me here. For macroscopic objects (those large compared to the scale where quantum-mechanical effects turn out to be significant) one finds that one path is overwhelmingly more probable than any other path, and all others can be ignored. That is the path that is predicted by the laws of classical mechanics, and this explains why the observed laws of motion of macroscopic objects are so well described by classical mechanics. But for particles moving on scales where quantum mechanics can make significantly different predictions than classical mechanics, several different paths may be equally probable. In this case, the final probability for a particle to go from A to B will depend on considering more than one possible path. Now this final probability that a particle that starts out at A will end up at B turns out to depend upon the sum of the individual "probabilities" for all the possible paths.

What makes quantum mechanics fundamentally different from classical mechanics is that these "probabilities" as calculated bear one fundamental difference from actual physical probabilities as we normally define them. While normal probabilities are always positive, the "probabilities" for the individual paths in quantum mechanics can be negative, or even "imaginary," that is, a number whose square is negative! (If you don't like to think about imaginary probabilities, you can get rid of this feature by imagining a world where "time" is an imaginary number. In this case, all the probabilities can be written as positive numbers. Hence the term *imaginary time.* No further explanation of this issue is necessary here, however. Imaginary time is merely a mathematical construct designed to help us handle the mathematics of quantum mechanics, and nothing more.) There is no problem calculating the final real physical probability for a particle to go from A to B, because after adding together the individual quantum-mechanical "probabilities" for each path, the laws of quantum mechanics then tell me to square the result in such a way that the actual physical probability is always a positive number.

The important point about all this is that I can consider the "probabilities" associated with two different paths, each of which, when taken separately, might yield a nonzero final probability but, when added together, cancel out to give a zero result. This is exactly what can happen for the electrons traveling to the phosphorescent screen through the two slits. If I consider a certain point on the screen, and I cover up one slit, I find that there is a nonzero probability for the electron to take a trajectory from A to B through this slit and get to the screen:

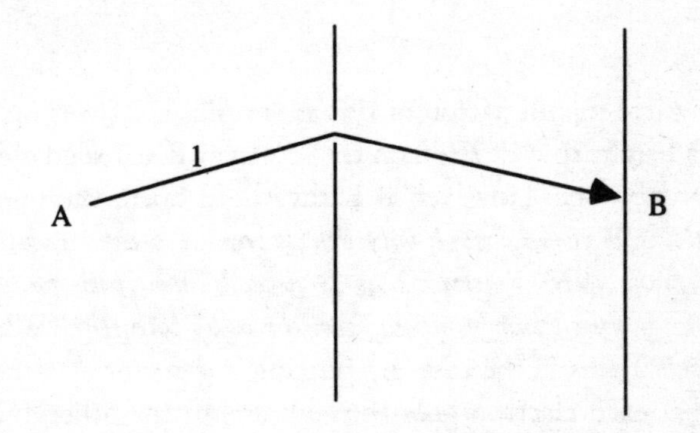

Similarly, there is a nonzero probability for it to go from A to B if I cover up the other slit:

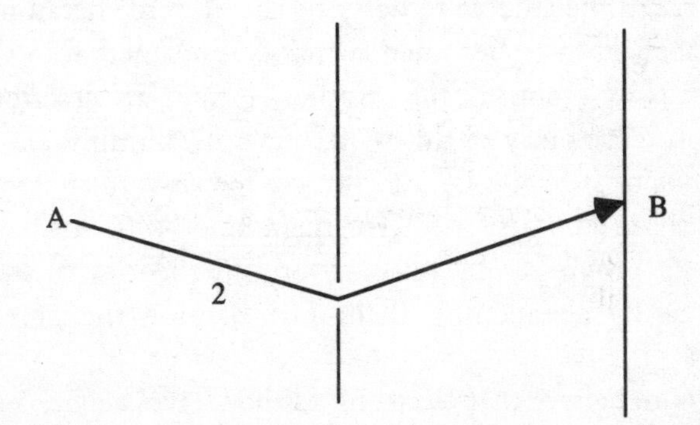

However, if I allow both possible paths, the final probability to go from A to B, based on the sum of the quantum-mechanical "probabilities" for each path, can be zero:

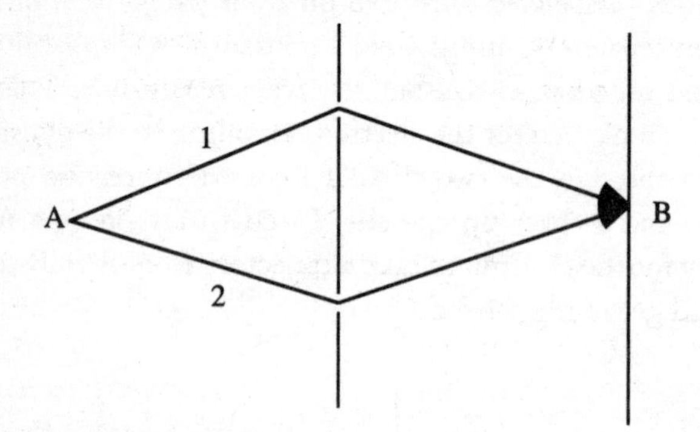

The physical manifestation of this is simple. If I cover up either slit, I see a bright spot develop on the screen at B as I send electrons through one by one. However, if I leave both open, the screen remains dark at B. *Even though only one electron at a time travels to the screen, the probability of arriving at B depends upon both paths being available, as if the electron somehow "goes through" both slits!* When we check to see if this is the case, by putting a detector at either slit, we find that each electron goes through one or the other, but now the spot at B is bright! Detecting the electron, forcing it to betray its presence, to make a choice if you will, has the same effect as closing one of the slits: It changes the rules!

I have taken the trouble to describe this in some detail not just to introduce you to a phenomenon that is fundamental to the way the world is at atomic scales. Rather, I want to describe how, hanging on tenaciously to this wild, but proven result, and to the implications of special relativity, we are forced into consequences that even those who first predicted them found difficult to accept. But physics progresses by pushing proven theories to their extremes, not by abandoning them just because the going gets tough.

If we are to believe that electrons, as they travel about, "explore" all trajectories available to them, impervious to our ability to check up on them, we must accept the possibility that some of these trajectories might be "impossible" if only they were measurable. In particular, since even repeated careful measurements of a particle's position at successive times cannot establish unambiguously its velocity between these times, by the uncertainty principle, it may be that for very short times a particle might travel faster than the speed of light. Now it is one of the fundamental consequences of the special relationship between space and time that Einstein proposed—in order, I remind you, to reconcile the required constancy of the speed of light for all observers—that nothing can be measured to travel faster than the speed of light.

We are now led to the famous question: If a tree falls in the forest and no one is there to hear it, does it make a sound? Or, perhaps more pertinent to this discussion: If an elementary particle whose average measured velocity is less than the speed of light momentarily travels faster than the speed of light during an interval so small that I cannot directly measure it, can this have any observable consequences? The answer is yes in both cases.

Special relativity so closely connects space and time that it constrains velocities, which associate a distance traveled with a certain time. It is an inevitable consequence of the new connections between space and time imposed by relativity that, were an object to be measured to travel faster than the speed of light by one set of observers, it could be measured by other observers to be going *backward* in time! This is one of the reasons such travel is forbidden (otherwise causality could be violated, and for example, as all science fiction writers are aware, such unacceptable possibilities could occur as shooting my grandmother before I was born!). Now quantum mechanics seems to imply that particles can, in principle, travel faster than light for intervals so small I cannot measure their velocity. As long as I cannot measure such velocities directly, there is no operational violation of special relativity. However, if the quantum theory is to remain consistent with special relativity, then during these intervals, even if I cannot measure them, such particles must be able to *behave* as though they were traveling backward in time.

What does this mean, practically? Well, let's draw the trajectory of the electron as seen by a hypothetical observer who might watch this momentary jump in time:

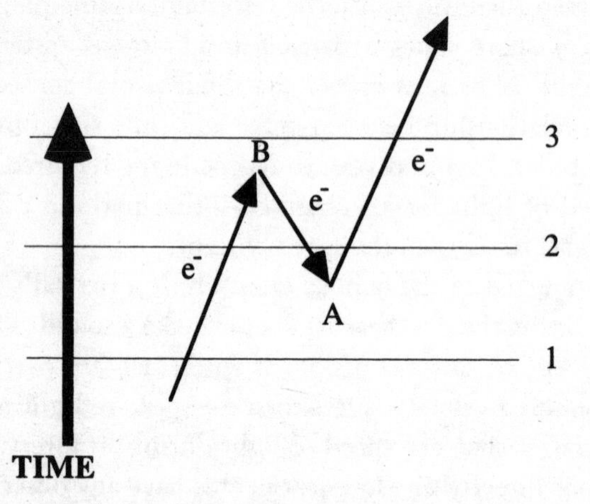

If such an observer were recording her observations at the three times labeled 1, 2, and 3, she would measure one particle at time 1, *three* particles at 2, and one particle again at 3. In other words, the number of particles present at any one time would not be fixed! Sometimes there may be one electron moving along its merry way, and at other times this electron may be accompanied by two others, albeit one of which apparently is moving backward in time.

But what does it mean to say an electron is moving backward in time? Well, I know a particle is an electron by measuring its mass and its electric charge, the latter of which is negative. The electron traveling from position B at one time to position A at an earlier time represents the flow of negative charge from left to right as I go backward in time. For an observer who is herself moving forward in time, as observers tend to do, this will be recorded as the flow of *positive* charge from right to left. Thus, our observer will indeed measure three particles present at times between 1 and 3, all of which will appear to move forward in time, but one of these particles, which will have the same mass as the electron, will have a positive charge. In this case, the series of events in the previous diagram would be pictured instead as:

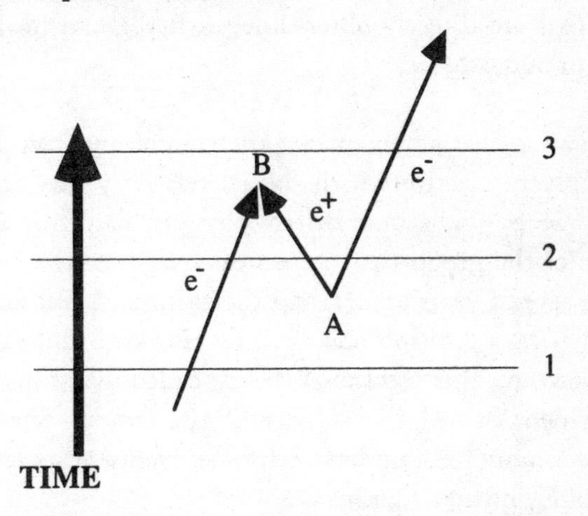

From this viewpoint, the picture is just a little less strange. At time 1, the observer sees an electron moving from left to right. At position A, at a time between 1 and 2, the observer suddenly sees an additional pair of particles appear from nothing. One of the particles, with positive charge, moves off to the left, and the other, with negative charge, moves off to the right. Some time later, at position B, the positive-charged particle and the original electron collide and disappear, leaving only the "new" electron continuing to move on its merry way from left to right.

Now, as I said, no observer can actually measure the velocity of the original electron for the time interval between 1 and 3, and thus no physical observer could directly measure this apparent spontaneous "creation" of particles from nothing, just as no observer will ever measure the original particle to be traveling greater than the speed of light. But whether or not we can measure it directly, if the laws of quantum mechanics allow for this possibility, they must, if they are to be made consistent with special relativity, allow for the spontaneous creation and annihilation of pairs of particles on time scales so short that we cannot measure their presence directly. We call such particles *virtual particles.* And as I described in chapter 1, processes such as those pictured above may not be directly observable, but they do leave an indirect signature on processes that are directly observable, as Bethe and his colleagues were able to calculate.

The equation that amalgamated the laws of quantum mechanics as they apply to electrons with special relativity was first written down in 1928 by the laconic British physicist Paul Adrian Maurice Dirac, one of the group that helped discover the laws of quantum mechanics several years earlier, and the eventual Lucasian Professor of Mathematics, a position now held by Hawking and earlier occupied by Newton. This combined theory, called quantum electrodynamics, which formed the subject of the famous Shelter Island meeting, was only fully understood some twenty years later, due to the work of Feynman and others.

No two physicists could have been more different than Dirac and Feynman. As much as Feynman was an extrovert, so much was Dirac an introvert. The middle child of a Swiss teacher of French in Bristol, England, young Paul was made to follow his father's rule to address him only in French, in order that the boy learn that language. Since Paul could not express himself well in French, he chose to remain silent, an inclination that would remain with him for the rest of his life. It is said (and may be true) that Niels Bohr, the most famous physicist of his day, and Director of the institute in Copenhagen where Dirac went to work after receiving his Ph.D. at Cambridge, went to visit Lord Rutherford, the British physicist, some time after Dirac's arrival. He complained about this new young researcher, who had not said anything since his arrival. Rutherford countered by telling Bohr a story along the following lines: A man walks into a store wanting to buy a parrot. The clerk shows him three birds. The first is a splendid yellow and white, and has a vocabulary of 300 words. When asked the price, the clerk replies, $5,000. The second bird was even more colorful than the first, and spoke well in four languages! Again the man asked the price, and was told that this bird could be purchased for $25,000. The man then spied the third bird, which was somewhat ragged, sitting in his cage. He asked the clerk how many foreign languages this bird could speak, and was told, "none." Feeling budget-conscious, the man expectantly asked how much this bird was. "$100,000" was the response. Incredulous, the man said, "What? This bird is nowhere near as colorful as the first, and nowhere near as conversant as the second. How on earth can you see fit to charge so much?" Whereupon the clerk smiled politely and replied, "This bird thinks!"

Be that as it may, Dirac was not one to do physics by visualization. He felt much more comfortable with equations, and it was only after playing with one set of equations for several years that he discovered a remarkable relation that correctly described the quantum-mechanical and special-relativistic behavior of electrons. It quickly became clear not only that this equation predicted the possible existence of a positive partner of the electron, which could exist as a "virtual" particle produced as part of a virtual pair of particles, but that this new object must also be able to exist on its own as a real particle in isolation.

At that time the only known positively charged particle in nature was the proton. Dirac and colleagues, who saw that his equation correctly predicted a number of otherwise unexplained features of atomic phenomena but did not want to depart too far from current orthodoxy, thus assumed that the proton must be the positive particle predicted by the theory. The only problem was that the proton was almost 2,000 times heavier than the electron, while the most naive interpretation of Dirac's theory was that the positive particle should have the same mass as the electron.

Here was an example where two known, well-measured theories of the physical world, when pushed to their limits, forced upon us paradoxical conclusions, just as relativity did for the unification of Galileo's ideas with electromagnetism. Yet, unlike Einstein, physicists in 1928 were not so ready to demand new phenomena to validate their ideas. It was not until 1932 that, quite by accident, the

American experimental physicist Carl Anderson, observing cosmic rays—the high-energy particles that continually bombard the Earth and whose origin ranges from nearby solar flares to exploding stars in distant galaxies—discovered an anomaly in his data. This anomaly could be explained only if there existed a new, positively charged particle whose mass was much closer to that of the electron than the proton. So it was that the "positron," the "antiparticle" of the electron predicted by Dirac's theory, was discovered. We now know that the same laws of quantum mechanics and relativity tell us that for every charged particle in nature, there should exist an antiparticle of equal mass and opposite electric charge.

Reflecting on his timidity in accepting the implications of his work amalgamating special relativity and quantum mechanics, Dirac is said to have made one of his rare utterances: "My equation was smarter than I was!" So, too, my purpose in relating this story is to illustrate again how the most remarkable results in physics often arise not from discarding current ideas and techniques, but rather by boldly pushing them as far as you can—and then having the courage to explore the implications.

I think I have pushed the idea of pushing ideas as far as they can be pushed about as far as I can push it. But by entitling this chapter "Creative Plagiarism," I don't mean just stretching old ideas to their limits; I also mean copying them over whole hog! Everywhere we turn nature keeps on repeating herself. For example, there are only four forces we know of in nature—the strong, weak, electromagnetic, and gravitational forces—and every one of them exists in the image of any of the others. Begin with Newton's Law of Gravity. The only other long-range force in nature, the force between charged particles, starts out as a direct copy of gravity. Change "mass" to "electric charge" and that's largely it. The classical picture of an electron orbiting a proton to make up the simplest atom, hydrogen, is *identical* to the picture of the moon orbiting the Earth. The strengths of the interactions are quite different, and that accounts for the difference of scale in the problem, but otherwise all of the results built up to describe the motion of the planets around the sun and apply in this case. We find out that the period of an electron's orbit around a proton is about 10^{-15} seconds, as opposed to one month for the moon around the Earth. Even this straightforward observation is enlightening. The frequency of visible light emitted by atoms is of the order of 10^{15} cycles/second, strongly suggesting that the electron orbiting around the atom has something to do with the emission of light, as is indeed the case.

Of course, there are important differences that make the electric force richer than gravity. Electric charge comes in two different types: positive and negative. Thus, electric forces can be repulsive as well as attractive. In addition, there is the fact that moving electric charges experience a magnetic force. As I described earlier, this leads to the existence of light, as an electromagnetic wave generated by moving charges. The theory of electromagnetism, in which all these phenomena are unified, then serves as a model for the weak interactions between particles in nuclei that are responsible for most nuclear reactions. The theories are so similar that it was eventually realized that they could be unified together into a single theory, which itself was a generalization of electromagnetism. The fourth force, the strong force between quarks that make up protons and neutrons, is also modeled on electromagnetism. This is reflected in its name, *quantum chromodynamics,* a descendant of quantum electrodynamics. Finally, with the experience gained from these theories we can go back to Newtonian gravity and generalize it and, lo and behold, we arrive at Einstein's general relativity. As the physicist Sheldon Glashow has said, physics, like the Ouroboros, the snake that eats its tail, returns full circle.

I want to end this chapter with a specific example that graphically demonstrates how strong the connections are between completely different areas of physics. Moreover, it is both topical and poetic. It has to do with the Superconducting Supercollider, now being built outside Waxahachi, Texas, for a total cost of about $8 billion. This sum of money is large enough to make the SSC topical; it is poetic because its name signifies in an unintended, subtle way the intellectual legacy on which it is built.

Anyone who has ever visited the site of a large particle physics laboratory has experienced the meaning of the words of the eminent physicist/educator Vicki Weisskopf, who has described these facilities as the gothic cathedrals of the twentieth century. In scale and complexity, they are to the twentieth century what the vast engineering church projects were to the eleventh and twelfth centuries (although I doubt they will last as long). The Superconducting Supercollider is planned to be 54 miles around, located over 100 feet below Texas farmland. Over 4,000 huge superconducting magnets will guide two streams of protons in opposite directions around the tunnel, causing them to collide together at energies 10 million times their rest mass. Each collision can produce on average over a thousand particles, and there can be tens of millions of collisions per second.

The purpose of this gargantuan machine is to attempt to discover the origin of "mass" in nature. We currently have no idea why elementary particles have the masses they do, why some are heavier than others, and why some, such as neutrinos, may have no mass at all. A number of strong theoretical arguments suggest that the key to this mystery can be probed at energies accessible at the SSC.

The name *Superconducting Supercollider* comes partly from the many superconducting magnets that make up its central "engine," magnets that without the aid of cooling to temperatures so low that the wires in them become superconducting, would otherwise be impossible or at least prohibitively expensive to build. To understand why the term is also appropriate on much deeper grounds, we have to go back some eighty years to a laboratory in Leiden, Holland, where the distinguished Dutch experimental physicist H. Kammerlingh Onnes discovered the amazing phenomenon we now call superconductivity. Onnes was cooling down the metal mercury to a low temperature in his laboratory in order to examine its properties. As you cool any material down, its resistance to the flow of electric current decreases, primarily because the motion of the atoms and molecules in the material that tends to block the flow of current decreases. However, when Onnes cooled the mercury down to 270° below zero (Celsius), he witnessed something unexpected: The electrical resistance vanished completely! I do not mean that there was *hardly* any resistance; I mean there was *none.* A current, once started, would continue to flow unchanged for very long periods in a coil of such material, even after the power source that had started the current flowing was removed. Onnes dramatically demonstrated this fact by carrying along with him a loop of superconducting wire containing a persistent current from his home in Leiden to Cambridge, England.

Superconductivity remained an intriguing mystery for almost half a century until a full microscopic theory explaining the phenomenon was developed in 1957 by the physicists John Bardeen, Leon Cooper, and J. Robert Schrieffer. Bardeen had already made an important contribution to modern science and technology by being the co-inventor of the transistor, the basis of all modern electronic equipment. The Nobel Prize in Physics that Bardeen shared with Cooper and Schrieffer in 1972 for their work on superconductivity was his second. (I was reading just the other day a letter of complaint to a physics magazine which made the ironic point that when Bardeen—the only person to win two Nobel Prizes in the same field and the co-inventor of a device that changed the way the world worked—died in 1992, it was hardly mentioned on TV. It would be nice if people were able to associate the pleasure they get from their transistor-driven stereos, TVs, games, and computers with the ideas that people like Bardeen produced.)

The key idea that led to a theory of superconductivity was in fact proposed by the physicist Fritz London in 1950. He suggested that this weird behavior was the result of quantum-mechanical phenomena, which normally affect behavior on only very small scales, suddenly extending to macroscopic scales. It was as if all the electrons in a conductor normally contributing to the current that flows when you attach this conductor to a power source were suddenly acting as a single, "coherent" configuration with a behavior governed more by the quantum-mechanical laws that control the individual electrons than by the classical laws that normally govern macroscopic objects. If all the electrons that conduct current act as a single configuration that stretches all the way across the conductor, then the flow of current cannot be thought of as being due to the motion of individual electrons that may bounce off obstacles as they move, producing a resistance to their motion. Rather, this coherent configuration, which spans the material, allows charge to be transported through it. It is as if in one state, this configuration corresponds to a whole bunch of electrons at rest. In another state, which is stable and time-independent, the configuration corresponds to a whole bunch of electrons that are moving uniformly.

This whole phenomenon can take place only because of an important property of quantum mechanics. Because the amount of energy that can be transferred to or from a finite-sized system occurs only in discrete amounts, or "quanta," the set of possible energy states for any particular particle in a finite system is reduced in quantum mechanics from a continuous set to a discrete set. This is because the particles can only change their energy by absorbing energy. But if energy can only be absorbed in discrete amounts, the set of possible energies the particles can have will also be discrete. Now, what happens if you have a whole bunch of particles in a box? If there are many possible different energy states for the particles, one might expect each of them to occupy a different discrete state, on average. Yet sometimes, under very special circumstances, it is possible that all of the particles might want to occupy a single state.

To understand how this might happen, consider the following familiar experience: You watch a comedy in a crowded movie theater and find it hilarious. You then rent the video to watch at home alone, and it is merely funny. The reason? Laughter is contagious. When someone next to you starts to laugh uproariously, it is difficult not to laugh along. And the more people who are laughing around you, the harder it is to keep from joining them.

The physical analogue of this phenomenon can be at work for the particles in the box. Say that in a certain configuration, two particles in the box can be attracted to each other, and thus lower their total energy by hanging out together. Once two particles are doing this, it may be even more energetically favorable for a third bystander particle to join the pack. Now, say this particular kind of attraction occurs only if the particles are in one out of all the possible configurations they can have. You can guess what will happen. If you start the particles out randomly, pretty soon they will all "condense" into the same quantum state. Thus, a coherent "condensate" is formed.

But there is more to it. Because the different quantum states in a system are separated into discrete levels, once all the particles in this system are condensed into a single state, there can be a sizable "gap" in total energy between this state and a state of the whole system where, say, one particle is moving around independently and the rest remain grouped. This is precisely the situation in a superconductor. Even though each electron is negatively charged, and therefore repels other electrons, inside the material there can be a small residual attraction between electrons due to the presence of all the atoms in the solid. This in turn can cause the electrons to pair together and then condense in a single, coherent quantum configuration. Now, say I connect this whole system to a battery. All of the electrons want to move together in the presence of the electric force. If any one of them were to scatter off an obstacle, retarding its motion, it would have to change its quantum state at the same time. But there is an "energy barrier" that prevents this, because the electron is so tightly coupled to all of its partners. Thus, the electrons all move together on their merry way, avoiding obstacles and producing no resistance.

Just by the remarkable behavior of this conglomeration of electrons, you can guess that there will be other properties of the material that are changed in this superconducting state. One of these properties is called the Meissner effect, after the German physicist W. Meissner, who discovered it in 1933. He found that if you put a superconductor near a magnet, the superconducting material will make every effort to "expel" the magnetic field due to the magnet. By this I mean that the electrons in the material will arrange themselves so that the magnetic field outside is completely canceled and remains zero inside the material. In order to do so, little magnetic fields must be created on the surface of the material to cancel the external magnetic field. Thus, if you bring the material near the north pole of a magnet, all sorts of little north poles will be created on the surface of the material to repel the initial field. This can be quite dramatic. If you take a material that isn't superconducting and put it on a magnet, it may just sit there. If you cool the whole system down so that the material becomes superconducting, it will suddenly rise and "levitate" above the magnet because of all the little magnetic fields created on the surface that repel the initial magnetic field.

There is another way to describe this phenomenon. Light, as I have indicated earlier, is nothing other than an electromagnetic wave. Jiggle a charge, and the changing electric and magnetic fields will result in a light wave that travels outward. The light wave travels at the speed of light because the dynamics of electromagnetism are such that the energy carried by the light wave cannot be considered to have any "mass" associated with it. Put another way, the microscopic quantum-mechanical objects that on small scales correspond to what we call an electromagnetic wave on macroscopic scales, called *photons,* have no mass.

The reason magnetic fields cannot enter a superconductor is because when the photons corresponding to this macroscopic field enter inside and travel through the background of all the electrons in their coherent state, the properties of these photons themselves change. They act as if they had a mass! The situation is similar to the way you act when you are roller-skating on a sidewalk, as opposed to roller-skating in the mud. The stickiness of the mud produces much more resistance to your motion. Thus, if someone were pushing you, you would act as if you were much "heavier" in the mud than on the sidewalk—in the sense that it would be much harder to push you. So, too, these photons find it much harder to propagate in the superconductor, because of their effective mass in this material. The result is that they don't get far, and the magnetic field doesn't permeate the material.

We are finally ready to talk about how all of this relates to the SSC. I said that it is hoped this machine will discover why all elementary particles have mass. Before reading the previous few pages, you might have thought that no two subjects could be more unrelated, but in fact it is likely that the solution to the elementary particle mystery is identical to the reason superconducting materials expel magnetic fields.

Earlier I said that electromagnetism served as a model for the force in nature that governs nuclear reactions such as those that power the sun, called the "weak" force. The reason is that the mathematical framework for the two forces is virtually identical, except for one important difference. The photon, which is the quantum entity corresponding to electromagnetic waves, which transmit electromagnetic forces, is massless. The particles that transmit the weak force, on the other hand, are not. It is for this reason that the weak force between protons and neutrons inside a nucleus is so short-range and that this force is never felt outside the nucleus, while electric and magnetic forces are felt at large distances.

Once this fact was recognized by physicists, it wasn't long before they began wondering what could be responsible for this difference. The same physics responsible for the weird behavior of superconductors suggests a possible answer. I have already described how the world of elementary-particle physics, where special relativity and quantum mechanics work together, has a weird behavior of its own. In particular, I argued that empty space need not really be empty. It can be populated by virtual-particle pairs, which spontaneously appear and then disappear, too quickly to be detected. I also described in chapter 1 how these virtual particles can have an effect on observed processes, like the Lamb shift.

Now it is time to put two and two together. If virtual particles can have subtle effects on physical processes, can they have a more dramatic effect on the properties of measured elementary particles?

Imagine that a new kind of particle exists in nature that has a close attraction to particles of the same type. If one pair of such particles burps into existence, as virtual particles are wont to do, it costs energy to do this, so the particles must disappear quickly if energy is not to be violated. However, if these particles are attracted to each other, it may be energetically favorable not just to pop a single pair out, but rather to pop two pairs. But if two pairs are better than one, why not three? And so on. It could be, if one arranges the attraction of these particles just right, that the net energy required of a coherent system of many such particles is, in fact, less than that in a system in which no such particles are around. In this case, what would happen? We would expect such a coherent state of particles spontaneously to generate itself in nature. "Empty" space would be filled with such a coherent background of particles in a special single quantum state.

What would be the effect of such a phenomenon? Well, we would not necessarily expect to observe the individual particles in the background, because to produce one real such particle moving on its own might require an incredible amount of energy, just as it costs energy to try to kick an electron out of its coherent configuration in a superconductor. Instead, as particles move amid this background, we might expect their properties to be affected.

If we were to arrange this background to interact with the particles that transmit the weak force, called the W and Z particles, and *not* with photons, then we might hope that this could result in the W and Z particles effectively acting as if they had a large mass. Thus, the real reason the weak force acts so differently from electromagnetism would be due not to any intrinsic difference but rather to this universal coherent background these particles move through.

This hypothetical analogy between what happens to magnetic fields in a superconductor and what determines the fundamental properties of "nature" might seem too fantastic to be true, except that it explains every experiment undertaken to date. In 1984, the

W and Z particles were discovered and since then have been investigated in detail. Their properties are in perfect agreement with what one would predict if these properties arose due to the mechanism I have described here.

What is next, then? Well, what about the masses of normal particles, such as protons and electrons? Can we hope to understand these, too, as resulting from their interactions with this uniform, coherent quantum state that fills up empty space? If so, the origin of all particle masses would be the same. How can we find out for sure? Simple: by creating the particles, called Higgs particles, after the Scottish physicist Peter Higgs, which are supposed to condense into empty space to get the whole game going. The central mission of the Superconducting Supercollider is usually said to be to discover the Higgs particle. The same theoretical predictions that reveal so well the properties of the W and Z particles tell us that the Higgs particle, if it exists, should have a mass within a factor of 10 or so of these particles, and this range is what the design of the SSC is chosen to explore.

I should say that this picture does not require the Higgs particle to be a fundamental elementary particle, like the electron or the proton. It could be that the Higgs is made up of pairs of other particles, like the pairs of electrons that themselves bind with one another to form a superconducting state in matter. And why does the Higgs exist, if it does? Is there a more fundamental theory that explains its existence, along with that of electrons, quarks, photons, and W and Z particles? We will be able to answer these questions only by performing experiments to find out.

I personally cannot see how anyone can fail to be awestruck at this striking duality between the physics of superconductivity and that which may explain the origin of all mass in the universe. But appreciating this incredible intellectual unification and wishing to pay to learn about it are two different things. It will, after all, cost up to $10 billion spread over ten years to build the SSC. The real issue of whether to build the SSC is not a scientific one—no properly informed individual can doubt the scientific worthiness of the project. It is a political question: Can we afford to make it a priority in a time of limited resources?

As I stressed at the beginning of this book, I believe that the major justification of the advances in knowledge likely at the SSC is cultural, not technological. We do not talk much about the plumbing facilities of ancient Greece, but we remember the philosophical and scientific ideals established there. These have filtered down through our popular culture, helping forge the institutions we use to govern ourselves and the methods we use to teach our young. The Higgs particle, if discovered, will not change our everyday life. But I am confident that the picture it is a part of will influence future generations, if only by exciting the curiosity of some young people, causing them perhaps to choose a career in science or technology. I am reminded here of a statement attributed to Robert Wilson, the first director of the large Fermilab facility currently housing the highest-energy accelerator in the world. When asked whether this facility would contribute to the national defense, he is said to have replied, "No, but it will help keep this country worth defending."

4

HIDDEN REALITIES

We shall not cease from exploration
And the end of all our exploring
Will be to arrive where we started
And know the place for the first time.
—T. S. Eliot, "Little Gidding," *Four Quartets*

YOU WAKE up one icy morning and look out the window. But you don't see anything familiar. The world is full of odd patterns. It takes a second for you to realize that you are seeing icicles on the window, which suddenly focus into place. The intricate patterns reflecting the sun's light then begin to captivate you.

Science museums call it the "aha" experience. Mystics probably have another name for it. This sudden rearrangement of the world, this new gestalt, when disparate data merge together to form a new pattern, causing you to see the same old things again in a totally new light, almost *defines* progress in physics. Each time we have peeled back another layer, we have discovered that what was hidden often masked an underlying simplicity. The usual signal? Things with no apparent connection can be recognized as one and the same.

The major developments in twentieth-century physics conform to this tradition, ranging from the fascinating discoveries of Einstein about space, time, and the universe, to the importance of describing how oatmeal boils. In discussing these "hidden realities," I don't want to get caught up in philosophical arguments about the ultimate nature of reality. This is the kind of discussion that tends to confirm my general view of philosophy, best expressed by the twentieth-century philosopher and logician Ludwig Wittgenstein: "Most propositions and questions that have been written about philosophical matters are not false, but senseless."[1]

Wondering, for instance, as Plato did, whether there is an external reality, with an existence independent of our ability to measure it, can make for long discussions and not much else. Having said this, I do want to use an idea that Plato developed in his famous cave allegory—in part because it helps me appear literate, but more important, because building upon it allows me to provide an allegory of my own.

Plato likened our place in the grand scheme of things to a person living in a cave, whose entire picture of reality comes from images—shadows cast on the wall—of the "real objects" that exist forever in the light of day, beyond the person's gaze. He argued that we, too, like the person in the cave, are condemned only to scratch the surface of reality through the confines of our senses.

One can imagine the difficulties inherent in the life of the prisoner of the cave. Shadows give at best a poor reflection of the world. Nevertheless, one can also imagine moments of inspiration. Say that every Sunday evening before the sun set, the following shadow was reflected on the wall:

And every Monday evening, the image below replaced it:

In spite of its remarkable resemblance to a cow, this is really an image of something else. Week in and week out the same images would be cast upon the wall, relentlessly changing and reappearing with clockwork regularity. Finally one Monday morning, awaking earlier than normal, our prisoner would also hear the sound of a truck combined with the clatter of metal. Being a woman of extraordinary imagination, combined with mathematical talent, she had a new vision suddenly pop into her head: These were not different objects, after all! They were one and the same. Adding a new dimension to her imagination, she could picture the actual extended object, a garbage can:

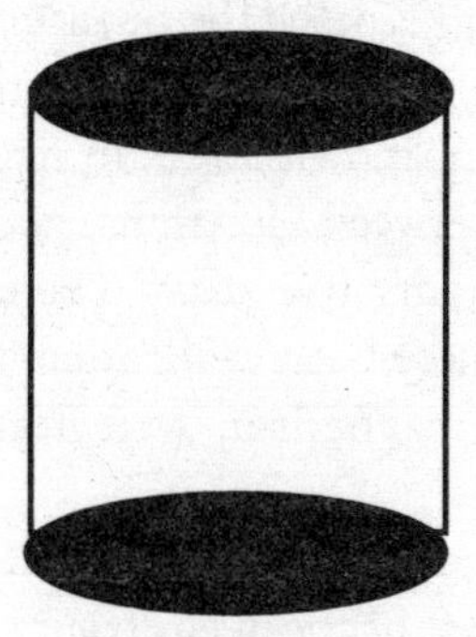

Standing upright on Sunday night, with the sun low in the sky directly behind it, it would cast the image of a rectangle. On Monday, after being tossed on its side by the collectors, it would cast the image of a circle. A three-dimensional object may, when viewed from different angles, cast very different two-dimensional projections. With this leap of inspiration, not only was a puzzle solved but different phenomena could now be understood to represent merely different reflections of the same thing.

Because of such realignments, physics does not progress as wheels within wheels; greater complexity does not always follow from finer detail. More often, new discoveries reflect sudden shifts in perception such as in the cave example. Sometimes the hidden realities that are exposed connect formerly disparate ideas, creating *less* from more. Sometimes, instead, they connect formerly disparate physical quantities and thus open up new areas of inquiry and understanding.

I have already introduced the unification that heralded the era of modern physics: James Clerk Maxwell's crowning intellectual achievement of the nineteenth century, the theory of electromagnetism and, with it, the "prediction" of the existence of light. It is appropriate that in the story of Genesis, light is created before anything else. It has also been the doorway to modern physics. The odd behavior of light caused Einstein to speculate about a new relationship between space and time. It also caused the founders of quantum mechanics to reinvent the rules of behavior at small scales, to accommodate the possibility that waves and particles might sometimes be the same thing. Finally, the quantum theory of light completed in the middle of this century formed the basis of our current understanding of all the known forces in nature, including the remarkable unification between electromagnetism and the weak interaction in the last twenty-five years. The understanding of light itself began with the more basic realization that two very different forces, electricity and magnetism, were really one and the same thing.

Earlier I sketched the discoveries of Faraday and Henry that established the connections between electricity and magnetism, but I don't think these give you as direct an idea of the depth or origin of such a connection as I would like. Instead, a thought experiment demonstrates directly how it is that electricity and magnetism are really different aspects of the same thing. As far as I know, this thought experiment was never actually performed prior to the experimental discoveries that led to the insights, but, with hindsight, it is very simple.

Thought experiments are an essential part of doing physics because they allow you to "witness" events from different perspectives at the same time. You may recall Akira Kurosawa's classic film *Rashomon,* in which a single event is viewed several different times and interpreted separately by each of the people present. Each different perspective gives us a new clue to intuit a broader, perhaps more objective, relationship among the events. Because it is impossible for one observer to have two vantage points at once, physicists take advantage of thought experiments of the type I will describe, following a tradition established by Galileo and brought to perfection under Einstein.

To perform this thought experiment, there are two facts you need to know. The first is that the only force a charged particle at rest feels, other than gravity, is an electric force. You can put the strongest magnet in the world next to such a particle and it will just sit there, oblivious. On the other hand, if you move a charged particle in the presence of a magnet, the particle will experience a force that changes its motion. This is called the Lorentz force, after the Dutch physicist Henrik Lorentz, who came close to formulating special relativity before Einstein. It has a most peculiar form. If a charged particle moves *horizontally* between the poles of a magnet, as shown below, the force on the particle will be upward, perpendicular to its original motion:

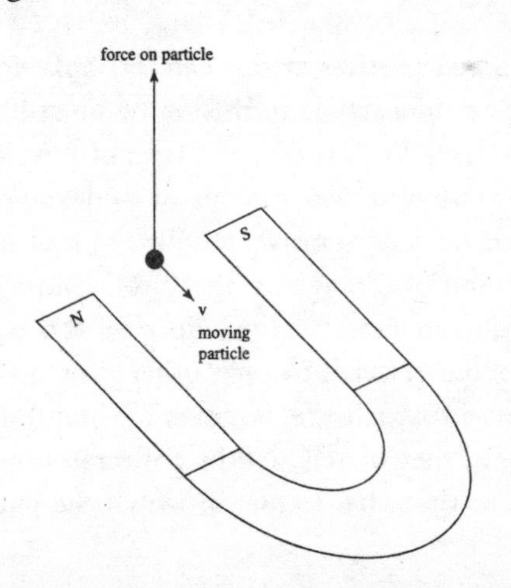

These two general features are sufficient to allow us demonstrate that an electric force to one person is a magnetic force to another. Electricity and magnetism are thus as closely related as the circle and rectangle on the cave wall. To see this, consider the particle in the previous diagram. If we are observing it in the laboratory, watching it move and get deflected, we know the force acting on it is the magnetic Lorentz force. But imagine instead that you are in a laboratory traveling at a constant velocity *along with* the particle. In this case, the particle will not be moving relative to you, but the magnet will be. You will instead see:

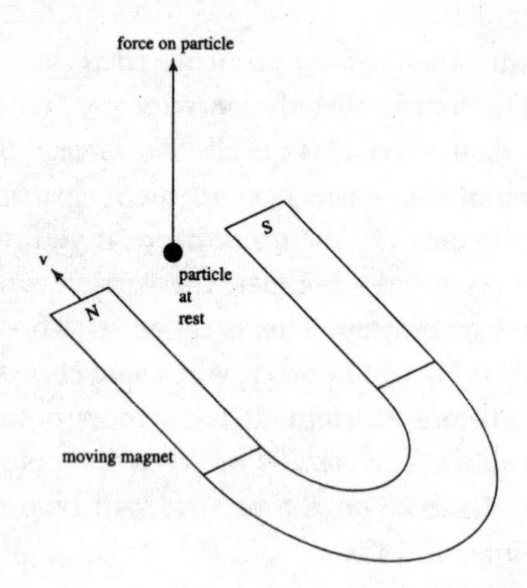

Because a charged particle at rest can feel only an electric force, the force acting on the particle in this frame must be electric. Also, since Galileo we have known that the laws of physics must appear the same for any two observers moving at a constant relative velocity. Thus, there is no way to prove, absolutely, that it is the particle that is moving and the magnets that are standing still, or vice versa. Rather, we can merely conclude that the particle and the magnets are moving relative to each other. But we have just seen that in the frame in which the magnets are standing still and the particle is moving, the particle will be deflected upward due to the magnetic force. In the other frame, in which the particle is at rest, this upward motion must be attributed to an electric force. As promised, one person's magnetic force is another's electric force. Electricity and magnetism are the different "shadows" of a single force, electromagnetism, as viewed from different vantage points, which depend upon your relative state of motion!

Next I want to jump to the present, to view how a much more recent development in physics—one that took place within the last twenty-five years and to which I alluded at the end of the last chapter—looks in this light. When I discussed the surprising relationship between superconductivity and the Supercollider, I described how one might understand the origin of mass itself as an "accident" of our restricted circumstances. Namely, we associate mass with only some particles because of the fact that we may be living amid a universal background "field" that preferentially "retards" their motion so as to make them appear heavy. I remind you that exactly the same thing happens for light in a superconductor. If we lived inside a superconductor, we would think that the carrier of light, the photon, was massive. Because we don't, we understand that the only reason light appears massive inside a superconductor is because of its interactions with the particular state of matter therein.

This is the trick. Stuck in a metaphorical cave, like a superconductor, how can we make the leap of inspiration to realize what actually exists outside our limited sphere of experience in a way that might unify otherwise diverse and apparently unrelated phenomena? I don't think there is any universal rule, but when we do make the leap, everything comes into focus so clearly that we know we have made the right one.

Such a leap began in the late 1950s, and ended in the early 1970s in particle physics. It slowly became clear that the theory that achieved completion after the discussions at Shelter Island, involving the quantum mechanics of electromagnetism, might also form the basis of the quantum theory of the other known forces in nature. As I earlier indicated, the mathematical frameworks behind both electromagnetism and the weak interactions responsible for most nuclear reactions are extremely similar. The only difference is that the particles that transmit the weak force are heavy, and the photon, the carrier of electromagnetism, is massless. In fact, it was shown in 1961 by Sheldon Glashow that these two different forces could in fact be unified into a single theory in which the electromagnetic force and the weak force were different manifestations of the same thing, but for the problem of the vast mass difference between the particles that transmit these forces, the photon and the W and Z particles.

Once it was recognized, however, that space itself could act like a vast superconductor, in the sense that a background "field" could effectively make particles "act" massive, it was soon proposed, in 1967, by Steven Weinberg and, independently, Abdus Salam, that this is exactly what happens to the W and Z particles, as I described at the end of the last chapter.

What is of interest here is not that a mechanism to give the W and Z particles mass had been discovered, but that in the absence of such a mechanism, it was now understood that the weak and electromagnetic forces are merely different manifestations of the same underyling physical theory. Once again, the observed considerable difference between two forces in nature is an accident of our situation. If we did not live in a space filled by the appropriate coherent state of particles, electromagnetism and the weak interactions would appear the same. Somehow it was managed, from disparate reflections on the wall, to discover the underlying unity present beyond the direct evidence of our senses.

In 1971, the Dutch physicist Gerard 't Hooft, then a graduate student, demonstrated that the mechanism proposed to give the W's and Z's mass was mathematically and physically consistent. In 1979, Glashow, Salam, and Weinberg were awarded the Nobel Prize for their theory, and in 1984, the particles that transmit the weak force, the W and Z particles, were discovered experimentally at the large accelerator at the Centre Européen pour Recherche Nucléaire (CERN) in Geneva, with their predicted masses.

This is not the only result of this new perspective. The success of viewing the weak and electromagnetic forces in a single framework that both mimics and extends the "simple" quantum theory of electromagnetism provided motivation to consider whether all the forces in nature could fall within this framework. The theory of the strong interactions, developed and confirmed after the theoretical discovery of asymptotic freedom I described in chapter 1, is of exactly the same general form, known as a "gauge" theory. Even this name has a history steeped in the notion of viewing different forces as different manifestations of the same underlying physics. Back in 1918, the physicist/mathematician Herman Weyl used one of the many similarities between gravity and electromagnetism to propose that they might be unified together into a single theory. He called the feature that related them a gauge symmetry—related to the fact that in general relativity, as we shall soon see, the *gauge,* or length scale, of local rulers used by different observers can be varied arbitrarily without affecting the underlying nature of the gravitational force. A mathematically similar change can be applied to the way different observers measure electric charge in the theory of electromagnetism. Weyl's proposal, which related classical electromagnetism and gravity, did not succeed in its original form. However, his mathematical rule turned out to play a vital role in the quantum theory of electromagnetism, and it is this property that is shared in common in the theories of the weak and strong interactions. It also turns out to be closely related to much of the current effort to develop a quantum theory of gravity and to unify it with the other known forces.

The "electroweak" theory, as it is now known, along with the theory of the strong interaction based on asymptotic freedom, have together become known as the Standard Model in particle physics. All existing experiments that have been performed in the last twenty years have been in perfect agreement with the predictions of these theories. All that remains to complete the unification of the weak and electromagnetic interactions in particular is to discover the exact nature of the coherent background quantum state that surrounds us and that we believe gives masses to the W's and Z's. We also want to know whether this same phenomenon is responsible for giving mass to all other observed particles in nature. This is what we hope the SSC will do for us.

Being a theoretical physicist, I am easily excited by these striking, if esoteric, realities hidden in the world of particle physics. Yet I know from conversations with my wife that these may seem too removed from everyday life for most people to get excited about. Nevertheless, they actually are directly tied to our own existence. If the known particles did not get exactly the masses they do, with the neutron being only one part in a thousand heavier than the proton, life as we know it would not have been possible. The fact that the proton is lighter than the neutron means that the proton is stable, at least on a time scale of the present age of the universe. Thus, hydrogen, made of a single proton and an electron and the most abundant element in the universe as well as being the fuel of stars such as our sun and the basis of organic molecules, is stable. Moreover, if the mass difference between the neutron and proton were different, this would have changed the sensitive equilibrium in the early universe that produced all the light elements we see. This combination of light elements contributed to the evolution of the first stars that formed, which not only contributed to the formation of our own sun some 5 billion to 10 billion years later but also produced all of the material inside our own bodies. It never ceases to amaze me that every atom in our own bodies originated in the fiery furnace of a distant exploding star! In this direct sense, we are all children of the universe. In a related vein, in the core of our sun, it is the mass difference between elementary particles that determines the rate of the energy-producing reactions there that fuel our own existence. Finally, it is the masses of these elementary particles that combine together to produce the readings on our bathroom scales that many of us dread stepping on.

As strong as the connection between the particle realm and our own may be, the progress of twentieth-century physics has not been confined to providing new perspectives just on phenomena beyond our direct perception. I want to return, in stages, from the extremes of scale I have been discussing to ones that are more familiar.

Nothing is more direct than our perception of space and time. It forms a crucial part of human cognitive development. Well known milestones in animal behavior are categorized by changes in spatial and temporal perception. For example, a kitten will walk unabashedly over a Plexiglas-covered hole until a certain age at which the animal begins to appreciate the dangerous significance of empty space beneath its feet. So it is all the more remarkable that we should discover, at the beginning of the twentieth-century, that space and time are intimately connected in a way that no one had previously suspected. Few would dispute that Albert Einstein's discovery of this connection through his special and general theories of relativity constitutes one of the preeminent intellectual achievements of our time. With hindsight, it is clear that his leaps were strikingly similar to those of our cave dweller.

As I have discussed, Einstein based his theory of relativity on the desire to maintain consistency with Maxwell's electromagnetic theory. In this latter theory, I remind you that the speed of light can be derived a priori in terms of two fundamental constants in nature: the strength of the electric force between two charges and the strength of the magnetic force between two magnets. Galilean relativity implies that these should be the same for any two observers traveling at a constant velocity relative to each other. But this would imply that all observers should measure the speed of light to be the same, regardless of their (constant) velocity or the velocity of the source of the light. Thus, Einstein arrived at his fundamental postulate of relativity: The speed of light in empty space is a universal constant, independent of the speed of the source or the observer.

In case the intuitive absurdity of this postulate has not hit home completely, let me give you another example of what it implies. In order to capture the White House in these times, it appears necessary for the winning political party to show that it encompasses the center, while the other party is either right or left of center. Some incredulity is therefore aroused when, before the election, both parties claim to have captured the center. Einstein's postulate, however, makes such a claim possible!

Imagine two observers in relative motion who pass by each other at the instant one of them turns on a light switch. A spherical shell of light will travel out in all directions to illuminate the night. Light travels so fast that we are normally unaware of its taking any time to move outward from the source, but it does. The observer, A, at rest with respect to the light bulb, would see the following shortly after turning on the light:

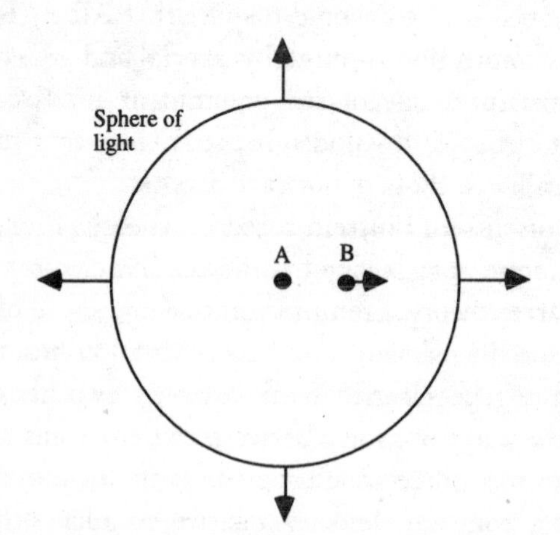

She would see herself at the center of the light sphere, and observer B, who is moving to the right relative to her, would have moved some distance in the time it took the light to spread out to its present position. Observer B, on the other hand, will measure these same light rays traveling outward as having the same fixed speed relative to him and thus as traveling the same distance outward relative to him, by Einstein's postulate. Thus, he will see *himself* as being at the center of the sphere, and A as having moved to the left of center:

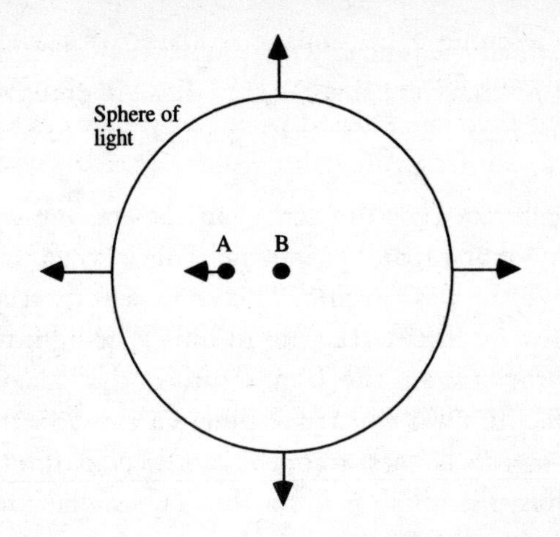

In other words, both observers will claim to be at the center of the sphere. Our intuition tells us this is impossible. But, unlike politics, in this case both protagonists actually *are* at the center! They must be if Einstein is right.

How can this be? Only if each observer somehow measures space and time differently. Then while one observer perceives that the distance between herself and all points on the sphere of light is the same and the distance between the other observer and the sphere is less in one direction and greater in another, the other observer can measure these same things and arrive at different answers. The absolutism of space-time has been traded for the absolutism of light's velocity. This is possible, and in fact all the paradoxes that relativity thrusts on us are possible, because our information about *remote* events is indirect. We cannot be both *here* and *there* at the same time. The only way we can find out what is happening *there* now is to receive some signal, such as a light ray. But if we receive it *now,* it was emitted *then.*

We are not accustomed to thinking this way, because the speed of light is so great that our intuition tells us that nearby events that we see *now* are actually happening *now.* But this is just an accident of our situation. Nevertheless, it is so pervasive an accident that had

Einstein not had the fundamental problems of electromagnetism to guide him in thinking about light, there is probably no way he would have been able to "see" beyond the cave reflection we call *now.*

When you take a picture with your camera, you are accustomed to imagining it as a snapshot in time: This is when the dog jumped on Lilli while she was dancing. This is not exactly true, however. It does represent an instant, but not in time. The light received at the camera at the instant the film recorded the image at different points on the film was emitted at different times, with those points displaying objects farthest from the camera emitting their light the earliest. Thus, the photo is not a slice in time but, rather, a set of slices in space at different times.

This "timelike" nature of space is normally not perceived because of the disparity between human spatial scales and the distances light can travel in human time scales. For example, in one-hundredth of a second, the length of time of a conventional snapshot, light travels approxmately 3,000 kilometers, or almost the distance across the United States! Nevertheless, even though no camera has such a depth of field, *now,* as captured in a photograph, is not absolute in any sense. It is unique to the observer taking the picture. It represents "here and now" *and* "there and then" to each different observer. Only those observers located at the same *here* can experience the same *now.*

Relativity tells us that, in fact, observers moving relative to one another *cannot* experience the same *now,* even if they are both *here* at the same instant. This is because their perceptions of what is "there and then" will differ. Let me give you an example. While I don't intend to rehash all the standard presentations in elementary texts on relativity, I will use one well-known example here because it is due to Einstein, and I have never seen a better one. Say that two observers are on two different trains, moving on parallel tracks at a constant velocity with respect to each other. It doesn't matter who is actually moving, because there is no way to tell, in an absolute sense, anyway. Say that at the instant these two observers, located in the center of their respective trains, pass each other, lightning strikes. Moreover, say that it strikes *twice* once at the front and once at the back of the trains. Consider the view of observer A at the instant he sees the light waves due to the flashes:

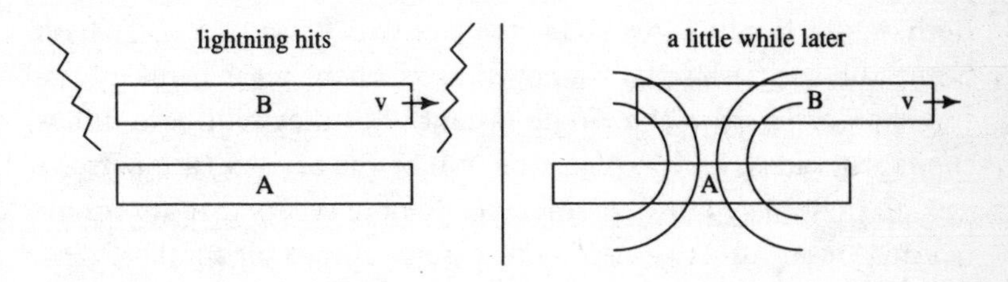

Since he observes the flashes simultaneously from either end of the train, and he is located in the center of his train, he will have no choice but to infer that the two flashes occurred at exactly the same time, which he could refer to as *now* although it was actually *then.* Moreover, since observer B is now to the right of A, B will see the flash from the right-hand lightning bolt before he sees the flash from the left-hand bolt.

This normally doesn't bother anyone, because we would all ascribe the difference in time of observation for B to the fact that he was moving toward one source of light and away from the other. However, Einstein tells us that there is no way for B to observe such an effect. The speed of both light rays toward him will be the same as if he were not moving at all. Thus, B will "observe" the following:

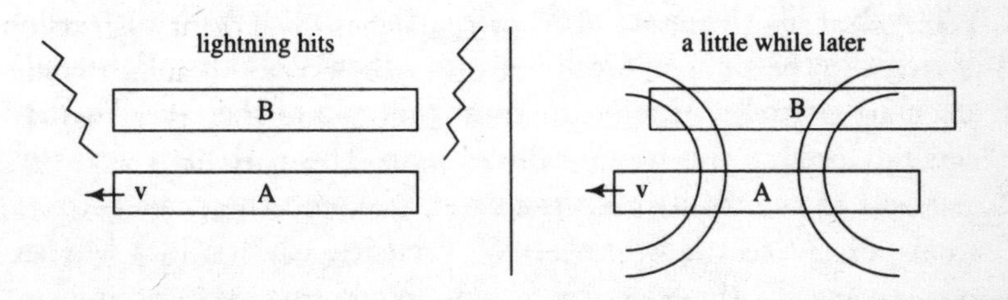

From this, B will infer—because (a) he sees one flash before the other, (b) the light was emitted from either end of a train in which he is sitting at the center, and (c) the light was traveling at the same speed in both directions—that the right-hand flash occurred before the left-hand flash. And for him, it actually did! There is no experiment either A or B can perform that will indicate otherwise. Both A and B will agree about the fact that B saw the right flash before the left flash (they cannot disagree about what happens at a single point in space at a single instant), but they will have different explanations. Each explanation will be the basis of each person's *now.* So these *nows* must be different. Remote events that are simultaneous for one observer need not be simultaneous for another.

The same kind of thought experiments led Einstein to demonstrate that two other facets of our picture of absolute space and absolute time must break down for observers moving relative to one another. A will "observe" B's clock to be running slowly, and B will "observe" A's clock to be running slowly. Moreover, A will "observe" B's train to be shorter than his own, and B will "observe" A's train to be shorter than his own.

Lest the reader think that these are merely paradoxes of perception, let me make it clear that they are not. Each observer will *measure* the passage of time to be different and will *measure* lengths to be different. Since, in physics, measurement determines reality, and we don't worry about realities that transcend measurement, this means that these things are *actually* happening. In fact, they are happening every day in ways that we can measure. The cosmic rays that bombard the Earth every second from space contain particles with very high energies, traveling very close to the speed of light. When they hit the upper atmosphere, they collide with the nuclei of atoms in the air and "break up" into a shower of other, lighter elementary particles. One of the most common of the particles that gets produced in this way is called a *muon.* This particle is virtually identical to the familiar electrons that make up the outer parts of atoms, except that it is heavier. We presently have no idea why an exact copy of the electron exists, prompting the prominent

American physicist I. I. Rabi to protest, "Who ordered that?" when the muon was discovered. In any case, the muon, because it is heavier than the electron, can decay into an electron and two neutrinos. We have measured the lifetime for this process in the laboratory and found that muons have a lifetime of about one-millionth of a second. A particle with a lifetime of one-millionth of a second traveling at the speed of light should go about 300 meters before decaying. Thus, muons produced in the upper atmosphere should never make it to the Earth. Yet they are the dominant form of cosmic rays (other than photons and electrons) that do!

Relativity explains this paradox. The muon's "clock" runs slowly compared to ours, because the muon is traveling near the speed of light. Therefore, in its own frame, the muon does decay on average in a few millionths of a second. However, depending upon how close to the speed of light the muon is traveling during this short time in its frame, perhaps several seconds could elapse in our frame on Earth. Time *really* does slow down for moving objects.

I can't resist leaving you with another paradox (my favorite!) that indicates how real these effects can be, while also underscoring how personal our notion of reality is. Say you have a brand-new, large American car that you want to show off by driving at a substantial fraction of the speed of light into my garage. Now your car, at rest, is 12 feet long. My garage is also just 12 feet long. If you are moving very fast, however, I will measure your car to be only, say, 8 feet long. Thus, there should be no problem fitting your car in my garage for a brief instant before either you hit the back wall or I open a door in the back of the garage to let you out. You, on the other hand, view me and my garage to be whizzing past you and thus to you my garage is only 8 feet long, while your car remains 12 feet long. Thus, it is impossible to fit your car in my garage.

The miraculous thing about this paradox is, once again, that both of us are right. I indeed can close my garage door on you and have you in my garage for a moment. You, on the other hand, will feel yourself hit the back wall before I close the door.

Nothing could be more real for either person, but as you can see, reality in this case is in the eye of the beholder. The point is that each person's *now* is subjective for distant events. The driver insists that *now* the front of his car is touching the back wall of the garage and the back is sticking out the front door, which is still open, while the garage owner insists that *now* the front door is closed and the front of the car has not yet reached the back wall.

If your *now* is not my *now* and your second is not my second, and your inch is not my inch, is there anything left to count on? The answer is yes, and it relates to the new connections between space and time that relativity uncovers. I have already described how once the finite velocity of light is taken into account, space has a timelike character. But these examples push that idea even further. One person's interval in space, such as the distance between the ends of a train measured at the same instant, can, for another person, also involve an interval in time. The second person will insist that these same measurements were in fact carried out at different times. Put another way, one person's "space" can be another person's "time."

In retrospect, this is not so surprising. The constancy of light connects space and time in a way in which they were not connected before. In order for a velocity—which is given by a distance traveled in a fixed time—to be measured as the same by two observers in relative motion, both space and time measurements must alter together between the two observers. There *is* an absolute, but it does not involve space separately, nor time separately. It must involve a combination of the two. And it is not hard to find out what this absolute is. The distance traveled by a light ray at a speed c for a time t is $d = ct$. If all other observers are to measure this same speed c for the light ray, then their times t' and distances d' must be such that $d' = ct'$. To write it in a fancier way, by squaring these expressions, the quantity $s^2 = c^2t^2 - d^2 = c^2 t'^2 - d'^2$ must equal zero and so must be the same for all observers. This is the key that can unlock our picture of space and time in exact analogy to our cave dweller's leap of insight.

Imagine shadows cast by a ruler on the cave wall. In one case, one may see this:

In another case, the same ruler can cast a shadow that looks like this:

To our poor cavewoman, fixed lengths will not be constant. What gives? We, who do not have to contend with two-dimensional projections but live in three-dimensional space, can see our way around the problem. Looking down on the ruler from above, we can see that in the two different cases the ruler was configured as shown below:

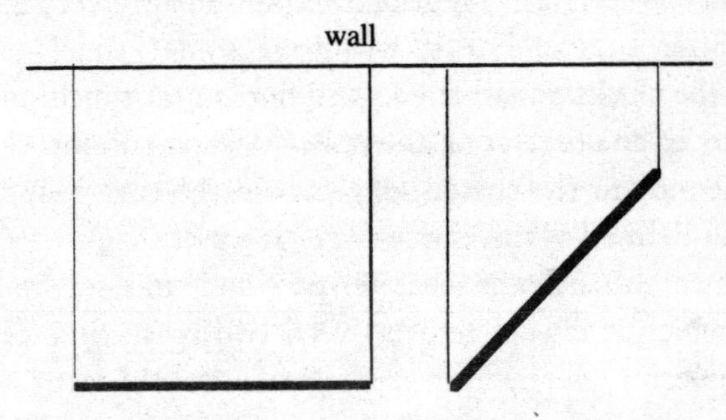

The second time, it had been rotated. We know that such a rotation does not change the ruler's length but merely changes the "component" of its length that gets projected on the wall. If, for example, there were two different observers viewing shadows projected at right angles, the rotated ruler would have projected lengths as shown:

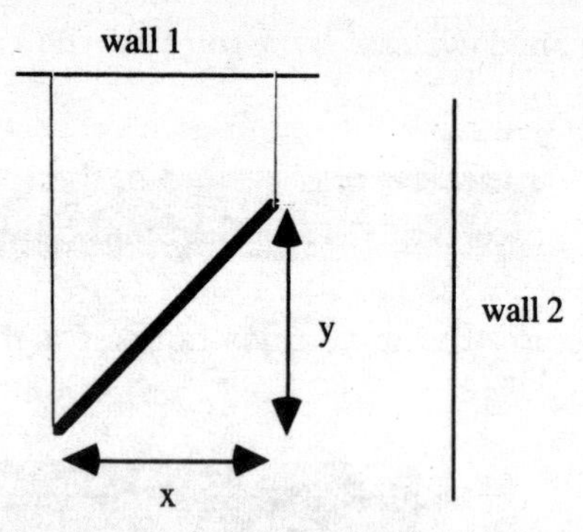

If the actual length of the ruler is L, Pythagoras tells us that $L^2 = x^2 + y^2$. Thus, even as the individual lengths, x and y, change for both observers, this combination will always remain constant. For one observer who measures the y direction, the original ruler would have zero extension, while for the other it would have maximal extension in the x direction. The rotated ruler, on the other hand, would have a nonzero y extension but smaller x extension.

This is remarkably similar to the behavior of space and time for two observers in relative motion. A train moving very fast may appear shorter to me, but it will have some "time" extension—namely, the clocks on either end will not appear synchronized to me if they are to an observer on the train. Most important, the quantity s is analogous to the spatial length L in the cave example. Recall that s was defined as the "space-time" interval $s^2 = c^2t^2 - d^2$. It represents a combination of separate space and time intervals between events, which is always zero between two space-time points lying on the trajectory of any light ray, regardless of the fact that different observers in relative motion will assign different individual values of d and t to the separate space and time intervals between the points they measure. It turns out that even if these points are not connected by a light ray but represent any two space-time points separated by length d and time t for one observer, the combination s (which need not be zero any longer) will be the same for all observers, again regardless of the fact that the separate length and time intervals measured may vary from observer to observer.

Thus, there *is* something about space and time that is absolute: the quantity s. This is for observers in relative motion what L is for observers rotated with respect to each other. It is the "space-time" length. The world we live in is thus best described as a four-dimensional space: The three dimensions of space and the "dimension" of time are coupled as closely (although not exactly in the same way) as the x and y directions were coupled above. And motion gives us different projections of this four-dimensional space on a three-dimensional slice we call *now,* just as rotations give different projections of a three-dimensional object onto the two-dimensional wall of the cave! Einstein, with his insights based on his insistence on the constancy of light, had the privilege of doing what many of us only dream about. He escaped from the confines of our cave to glimpse for the first time a hidden reality beyond our limited human condition, just as did our cave dweller who discovered that a circle and a rectangle were really reflections of a single object.

To his credit, Einstein did not stop there. The picture was not yet complete. Again he used light as his guide. All observers moving at constant relative velocities observe a light ray to have the same velocity c relative to them, and thus not one of them can prove that it is he who is standing still and the others who are moving. Motion is relative. But what about when they aren't moving at a constant velocity? What if one of them is accelerating? Will everyone in this case unambiguously agree that the odd one out is accelerating, including this lone observer? To gain some insight into these issues, Einstein considered an experience we have all had. When you are in an elevator, how do you know when and in what direction it has started to move? Well, if it begins to move upward, you momentarily feel heavier; if it moves downward, you momentarily feel lighter. But how do you *know* that it is actually moving, and not that gravity has suddenly gotten stronger?

The answer is, you don't. There is not a single experiment you can do in a closed elevator that can tell you whether you are accelerating or in a stronger gravitational field. We can make it even simpler. Put the elevator in empty space, with no Earth around. When the elevator is at rest, or moving at a constant velocity, nothing pulls you toward the floor. If the elevator is accelerating upward, however, the floor must push upward on you with some force to accelerate you along with the elevator. You in the elevator will feel yourself pushing down on the floor. If you have a ball in your hand and let it go, it will "fall" toward the floor. Why? Because if it were initially at rest, it would want to stay that way, by Galileo's Law. The floor, however, would be accelerating upward, toward it. From your vantage point, which is accelerating upward along with the elevator, the ball would fall down. What's more, this argument is independent of the mass of the ball. If you had six balls, all with different masses, they would all "fall" with the same acceleration. Again, this is because the floor would actually be accelerating upward toward them all at the same rate.

If Galileo was along with you in the elevator, he would swear he was back on Earth. Everything he spent his career proving about the way objects behave at the Earth's surface would be the same for objects in the elevator. So, while Galileo realized that the laws of physics should be identical for all observers moving with a constant velocity, Einstein realized that the laws of physics are identical for observers moving at a constant acceleration as those in a constant gravitational field. In this way, he argued that even acceleration is relative. One person's acceleration is another's gravity.

Again, Einstein peered beyond the cave. If gravity can be "created" in an elevator, maybe we are *all* just in a metaphorical elevator. Maybe what we call gravity is really related to our particular vantage point. But what is particular about our vantage point? We are on the Earth, a large mass. Perhaps what we view as a force between us and the Earth's mass can be viewed instead as resulting from something that the presence of this mass does to our surroundings, to space and time.

To resolve the puzzle, Einstein went back to light. He had just shown that it is the constancy of light that determines how space and time are woven together. What would a light ray do in the elevator accelerating in space? Well, to an outside observer, it would go in a straight line at a constant velocity. But inside the elevator, which is accelerating upward during this time, the path of the light ray would appear to look like this:

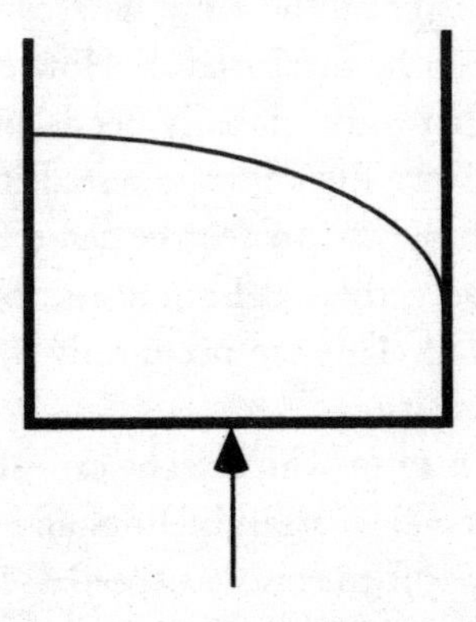

In the frame of the elevator, the light would appear to bend downward, because the elevator is accelerating upward away from it! In other words, it would appear to fall. So, if our accelerating elevator must be the same as an elevator at rest in a gravitational field, light must also bend in a gravitational field! This is actually not so surprising. Einstein had already shown that mass and energy were equivalent and interchangeable. The energy of a light ray will increase the mass of an object that absorbs it. Similarly, the mass of an object that emits a light ray decreases by an amount proportional to the energy of the radiation. Thus, if light can carry energy, it can act as though it has mass. And all massive objects fall in a gravitational field.

There is a fundamental problem with this notion, however. A ball that is falling, speeds up! Its velocity changes with position. However, the constancy of the speed of light is the pedestal on which special relativity is built. It is a fundamental tenet of special relativity that light travels at a constant velocity for all observers, regardless of their velocity relative to the light ray as seen by someone else. Thus, an observer at the top left-hand side of the elevator observing the light enter would be expected to measure the speed of light as *c,* but so would an observer at the lower right, measuring the light as it exits the elevator. It does not matter that the bottom observer is moving faster at the time he views the light than the first observer was when he measured it. How can we reconcile this result with the fact that light globally bends and therefore must be "falling"? Moreover, since Einstein suggested that if I am in a gravitational field I am supposed to see the same thing as if I were in the accelerating elevator, then, if I am *at rest,* but in a gravitational field, light will also fall. This can occur only if the velocity of light globally varies with position.

There is only one way in which light can globally bend and accelerate but locally travel in straight lines and always be measured by observers at each point to travel at speed *c.* The rulers and clocks of different observers, *now in a single frame*—that of the accelerating elevator, or one at rest in a gravitational field—must vary with position!

What happens to the global meaning of space and time if such a thing happens? Well, we can get a good idea by returning to our cave. Consider the following picture, showing the trajectory of an airplane from New York to Bombay as projected on the flat wall of the cave:

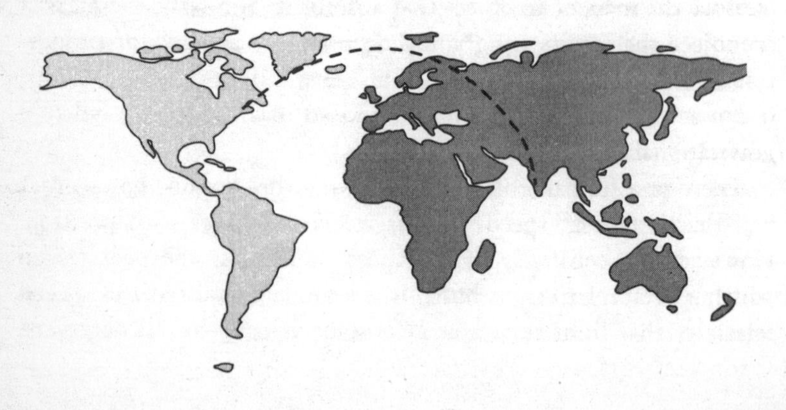

How can we make this curved trajectory look locally like a straight line, with the airplane moving at a constant velocity? One way would be to allow the length of rulers to vary as one moved across the surface, so that Greenland, which looks much larger in this picture than all of Europe, is in fact measured to be smaller by an observer who first measures the size of Greenland while he is there and then goes and measures the size of Europe with the same ruler when he is there.

It may seem crazy to propose such a solution, at least to the person in the cave. But we know better. This solution is really equivalent to recognizing that the underlying surface on which the picture is drawn is, in fact, *curved.* The surface is really a sphere, which has been projected onto a flat plane. In this way, distances near the pole are stretched out in the projection compared to the actual distances measured on the Earth's surface. Seeing it as a sphere, which we can do by benefit of our three-dimensional perspective, we are liberated. The trajectory shown above is really that of a line of longitude, which is a straight line drawn on a sphere and is the shortest distance between two points. An airplane traveling at a constant velocity in a straight path between these points would trace out such a curve.

What conclusion can we draw from this? If we are to be consistent, we must recognize that the rules we have found for an accelerating frame, or for one in which a gravitational field exists, are not unreasonable but are equivalent to requiring the underlying space-time to be *curved!* Why can't we sense this curvature directly if it exists? Because we always get a local view of space from the perspective of one point. It is just as if we were a bug living in Kansas. His world, which consists of the two-dimensional surface he crawls on, seems flat as a board. It is only by allowing ourselves the luxury of embedding this surface in a three-dimensional framework that we can directly picture our orb. Similarly, if we wanted to picture directly the curvature of a *three*-dimensional space, we would have to embed it in a four-dimensional framework, which is as impossible for us to picture as it would be for the bug whose life is tied to the Earth's surface, and for whom three-dimensional space is beyond his direct experience.

In this sense, Einstein was the Christopher Columbus of the twentieth century. Columbus argued that the Earth was a sphere. In order to grasp that hidden reality, he argued that he could set sail to the west and return from the east. Einstein, on the other hand, argued that to see that our three-dimensional space could be curved, one merely had to follow through on the behavior of a light ray in a gravitational field. This allowed him to propose three classic tests of his hypothesis. First, light should bend when it travels near the sun twice as much as if it were merely "falling" in a flat space. Second, the elliptical orbit around the sun of the planet Mercury would shift in orientation, or precess, by a very small amount each year, due to the small curvature of space near the sun. Third, clocks would run faster at the bottom of a tall building than at the top.

The orbit of Mercury had long been known to precess, and it turned out that this rate was exactly that calculated by Einstein. Nevertheless, explaining something that is already seen is not as exciting as predicting something completely new. Einstein's two other predictions of general relativity fell in this latter category. In 1919, an expedition led by Sir Arthur Stanley Eddington to South America to observe a total solar eclipse reported that the stars near the sun that could be observed during the darkness were shifted from where they were otherwise expected to be by exactly the amount Einstein had predicted. Light rays thus appeared to follow curved trajectories near the sun, and Einstein became a household name! It was not until forty years later that the third classical test proposed by Einstein was performed, in the basement of the Harvard physics laboratory of all places. Robert Pound and George Rebka demonstrated that the frequency of light produced in the basement was measured to be different when this light beam was received on top of the building. The frequency shift, though extremely small, was again precisely that by which Einstein had predicted a clock ticking at such a frequency would be shifted at the two different heights.

From the point of view of general relativity, the curved and accelerated trajectories followed by objects in a gravitational field, including light, can be thought of as artifacts of the underlying curvature of space. Again, a two-dimensional analogy is useful. Consider the two-dimensional projection of the motion of an object spiraling in toward larger object, as seen on the cave wall:

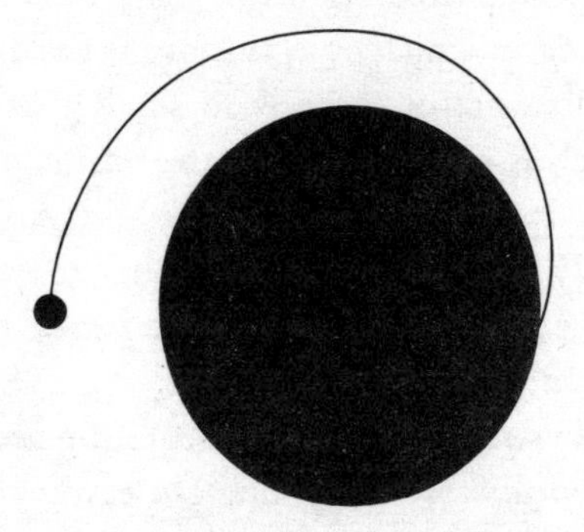

We could postulate a force between the objects to explain this. Or we could imagine that the actual surface on which this motion occurs is curved, as we can picture by embedding it in three dimensions. The object feels no force exerted by the larger ball, but rather follows a straight trajectory on this curved surface:

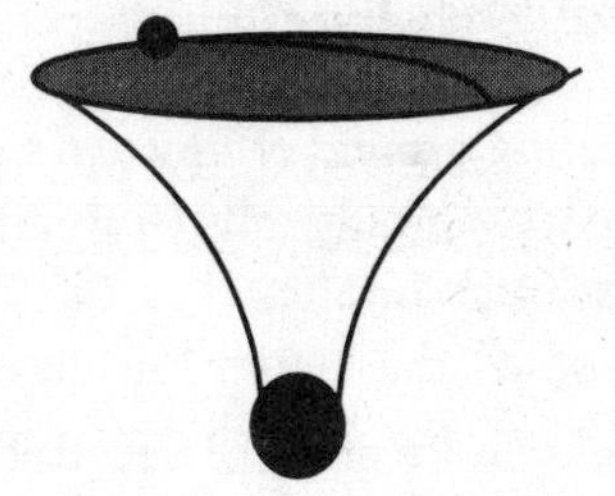

In just such a way, Einstein argued that the force of gravity that we measure between massive objects can be thought of instead as a consequence of the fact that the presence of mass produces a curvature of space nearby, and that objects that travel in this space merely follow straight lines in curved space-time, producing curved trajectories. There is thus a remarkable feedback between the presence of matter and the curvature of space-time, which is again reminiscent of the Ouroboros, the snake that eats itself. The curvature of space governs the motion of the matter, whose subsequent configuration in turn governs the curvature of space. It is this feedback between matter and curvature that makes general relativity so much more complicated to deal with than Newtonian gravity, where the background in which objects move is fixed.

Normally the curvature of space is so small that its effects are imperceptible, which is one reason why the notion of curved space seems foreign to us. In traveling from New York to Los Angeles, a light ray bends only about 1 millimeter due to the curvature of space induced by the Earth's mass. There are times, however, when even small effects can add up. For example, the 1987 Supernova I referred to previously was one of the most exciting astronomical observations of this century. However, one can easily calculate—and, in fact, a colleague and I were so surprised by the result that we wrote a research paper on its importance—that the small curvature through which the light from the 1987 Supernova passed as it traveled from one end of our galaxy to the other to reach us was sufficient to delay its arrival by up to nine months! Had it not been for general relativity and the curvature of space, the 1987 Supernova would have been witnessed in 1986!

The ultimate testing ground of Einstein's ideas is the universe itself. General relativity not only tells us about the curvature of space around local masses but implies that the geometry of the entire universe is governed by the matter within it. If there is sufficient average density of mass, the average curvature of space will be large enough so that space will eventually curve back on itself in the three-dimensional analogue of the two-dimensional surface of a sphere. More important, it turns out that in this case the universe would have to stop expanding eventually and recollapse in a "big crunch," the reverse of the big bang.

There is something fascinating about a "closed" universe, as such a high-density universe is called. I remember first learning about it as a high school student when I heard the astrophysicist Thomas Gold lecture, and it has stayed with me ever since. In a universe that closes back upon itself, light rays—which, of course, travel in straight lines in such a space—will eventually return back to where they started, just like latitudes and longitudes on the surface of the Earth. Thus, light can never escape out to infinity. Now, when such a thing happens on a smaller scale, that is, when an object has such a high density that not even light can escape from its surface, we call it a black hole. If our universe is closed, we are actually living inside a black hole! Not at all like you thought it would be, or like the Disney movie! That is because, as systems get larger and larger, the average density needed to produce a black hole gets smaller and smaller. A black hole with the mass of the sun would be about a kilometer in size when it first formed, with an average density of many tons per cubic centimeter. A black hole with the mass of our visible universe, however, would first form with a size comparable to the visible universe and an average density of only about 10^{-29} grams per cubic centimeter!

The current wisdom, though, suggests that we do not live inside a black hole. We believe that the average density of matter in space is such that we live in a universe that is just on the borderline between a closed universe, which closes in on itself and which will eventually recollapse, and an open universe, which is infinite and will continue to expand unabated forever. This borderline case, called a "flat" universe, is also infinite in spatial extent and will expand forever, with the expansion continuing to slow down but never quite stopping. Even a flat universe requires much more matter than we can account for with visible material—about 100 times as much, in fact. It is because we believe the universe is flat that we think that 99 percent of the universe must be made of dark matter, invisible to telescopes.

How can we tell if this supposition is correct? One way is to try to determine the total density of matter around galaxies and clusters of galaxies, as I described in chapter 3. There is another way, at least in principle, which is essentially the same as an intelligent bug in Kansas trying to determine whether the Earth is round without going around it and without leaving the surface. Even if such a bug could not picture a sphere in his mind, just as we cannot picture a curved three-dimensional space in ours, he might envisage it by generalizing a flat two-dimensional experience. There are geometrical measurements that can be done at the Earth's surface that are consistent only if this surface is a sphere. For example, Euclid told us over twenty centuries ago that the sum of the three angles in any triangle drawn on a piece of paper is 180°. If I draw a triangle with one 90° angle, the sum of the other two angles must be 90°. So each of these angles must be less than 90°, as shown in the two cases below:

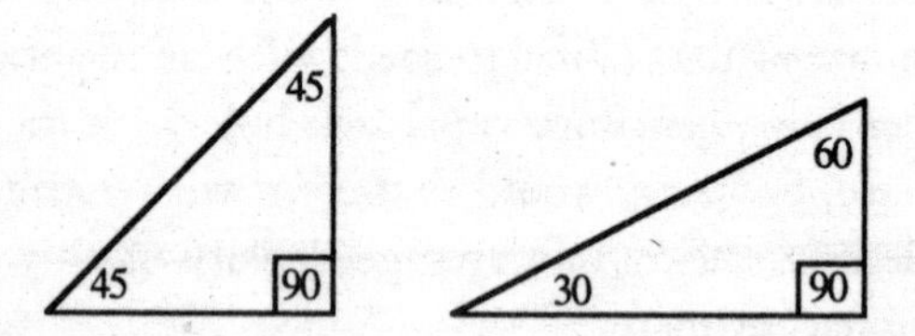

This is true only on a flat piece of paper, however. On the surface of the sphere, I can draw a triangle for which *each* of the angles is 90°! Simply draw a line along the equator, then go up a longitude line to the North Pole, and then make a 90° angle and go back down another longitude to the equator:

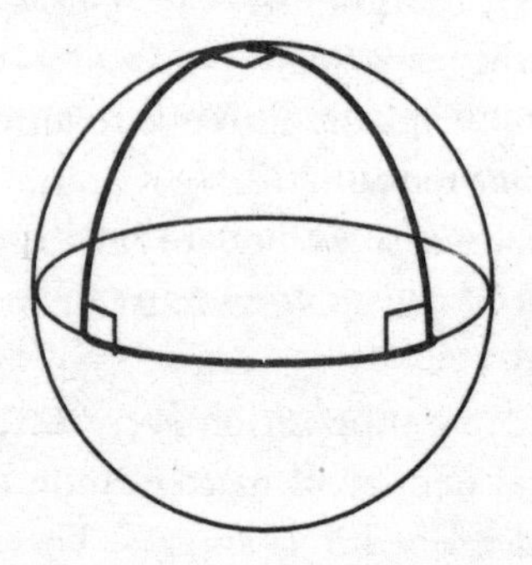

Similarly, you may remember that the circumference of a circle of radius *r* is $2\pi r$. However, on a sphere, if you travel out a distance *r* in all directions from, say, the North Pole, and then draw a circle connecting these points, you will find that the circumference of this circle is smaller than $2\pi r$. This is easier to understand if you look at the sphere from outside:

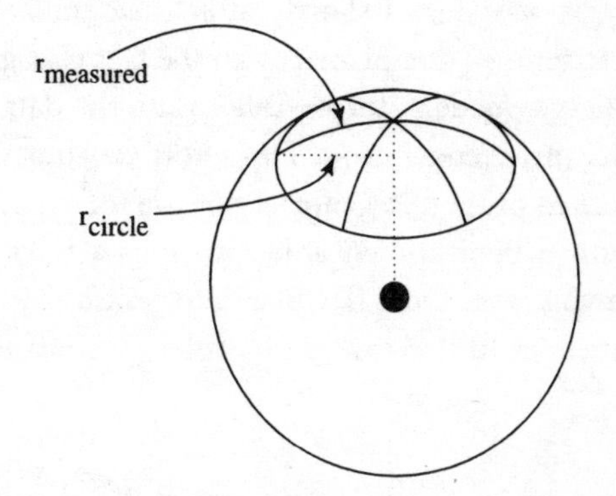

If we were to lay out big triangles, or big circles, on the Earth's surface, we could use the deviations from Euclid's predictions to measure its curvature and we would find that it is a sphere. As can be seen from the drawings, however, in order to deviate significantly from the flat-surface predictions of Euclid, we would have to produce extremely large figures, comparable to the size of the Earth. We can similarly probe the geometry of our three-dimensional space. Instead of using the circumference of circles, which is a good way to map out the curvature of a two-dimensional surface, we could instead use the area or volume of spheres. If we consider a large enough sphere of radius *r* centered on the position of the Earth, the volume inside this sphere should deviate from the prediction of Euclid if our three-dimensional space is curved.

How can we measure the volume of a sphere whose size is a significant fraction of the visible universe? Well, if we assume that the density of galaxies in the universe is roughly constant over space at any time, then we can assume that the volume of any region will be directly proportional to the total number of galaxies inside that region. All we have to do, in principle, is *count* galaxies as a function of distance. If space were curved, then we should be able to detect a deviation from Euclid's prediction. This was, in fact, tried in 1986 by two young astronomers who were then at Princeton, E. Loh and E. Spillar, and the results they announced purported to give evidence for the first time that the universe is flat, as we theorists expect it is. Unfortunately, it was shown shortly after they announced their results that uncertainties, due primarily to the fact that galaxies evolve in time and merge, made it impossible to use the data then at hand to draw any definitive conclusion. This effort continues.

Another way to probe the geometry of the universe is to measure the angle spanned by a known object such as a ruler held at some distance from your eye. On a flat plane, for example, it is clear that this angle continues to decrease as the ruler gets farther and farther away:

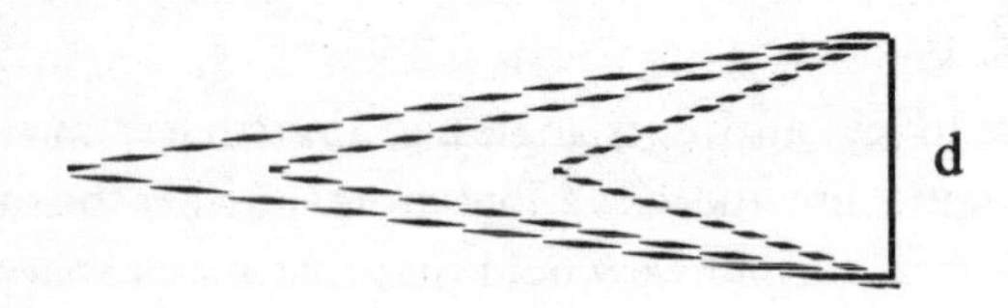

However, on a sphere, this need not be the case:

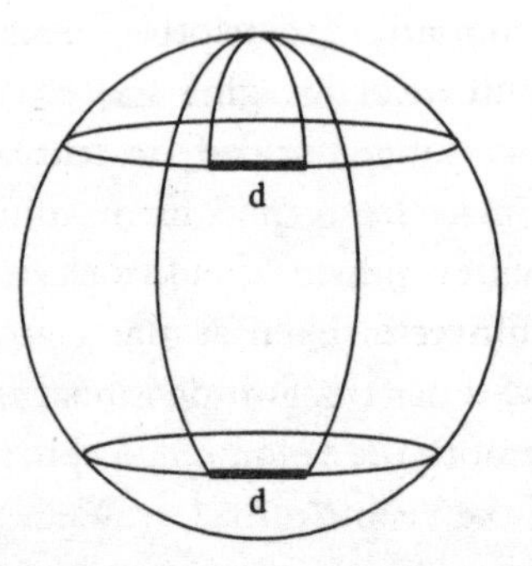

Recently, a systematic study was performed of the angle spanned by very compact objects in the center of distant galaxies measured with radio telescopes out to distances of almost half the size of the visible universe. The behavior of this angle with increasing distance is again almost exactly what one would predict for a flat universe. A colleague of mine and I have shown that this test too has possible ambiguities due to the possible cosmic evolution over time of the objects examined. Further tests will be needed to resolve this issue completely.

Whatever the accuracy of the present geometrical tests of cosmology, it is remarkable how far our understanding of space and time has developed since Einstein first unearthed the hidden connections between them. We now understand that we live in a four-dimensional universe, in which each observer must define his or her own *now* and, in so doing, partition space-time into the separate entities we perceive as space and time. We understand that space and time are also inextricably linked to matter, and that the curvature of space-time near massive objects is what we perceive as gravity. And we are on the brink of measuring the curvature of our universe and, from that, knowing what lies in store for the future of the universe. We may live in a metaphorical cave, but the shadows on the wall have so far provided unerring evidence that what they momentarily mask are remarkable connections that make the universe more comprehensible, and more predictable.

Before I wax too profound, I want to arrive back at the world of everyday phenomena to end this chapter. I promised examples that are close to home, and even though I started out with something simple like space and time, I ended up talking about the whole universe. But it is not just at the forefront of microscopic and macroscopic scales where hidden connections lie in wait to simplify our picture of the universe. Even as the grand discoveries about space, time, and matter that I have described in this chapter were being made, new connections have been unearthed about the nature of materials as exotic as oatmeal, and as diverse as water and iron. As I shall describe in the final chapter, while the subject of these discoveries is everyday physics, the ramifications include changing the way we picture the search for the "ultimate" theory.

Everyday materials appear extremely complicated. They must be because they differ widely in their behavior. One of the reasons why chemical engineering and materials science are such rich fields, and why industry supports significant materials research, is because substances can be designed to satisfy almost any property that might be required. Sometimes new materials are developed by accident. High-Temperature Superconductivity, for example, a subject of intense current interest, began almost as an alchemic accident by two researchers at IBM laboratories who were exploring a new set of materials in hopes of finding a new superconductor, but with no solid theoretical reasons behind their approach. On the other hand, more often than not, materials are developed based on a combination of empirical expertise and theoretical guidance. Silicon, for example, the main component in the semiconductor devices that run our computers (and our lives), has spawned a research field to search for materials with properties that might be more amenable to certain semiconductor applications. One such material is gallium, which has been stockpiled because of its assumed utility in the next generation of semiconductors.

Even the simplest and most commonplace materials have exotic behaviors. I will always remember my high school physics teacher telling me, tongue in cheek, of course, that there are two things in physics that prove the existence of God. First, water, almost alone of all materials, expands when it freezes. If this rare characteristic were not so, lakes would freeze from the bottom up instead of the top down. Fish would not survive the winter, and life as we know it would probably not have developed. Next, he pointed to the fact that the "coefficient of expansion"—that is, the amount by which a material expands when heated—of concrete and steel are virtually identical. If this were not the case, then modern large buildings would not be possible because they would buckle in the summer and winter. (I must admit that I found the second example a little less compelling, because I am sure that if concrete and steel did not have the same coefficient of expansion, some building materials could have been developed that did.)

Back to the first example, the fact that water, perhaps the most common substance on Earth, reacts differently than most other materials when it freezes is interesting. In fact, aside from the fact that water expands when it freezes, it provides in every other sense a paradigm for how materials change as external physical conditions change. At temperatures that exist on Earth, water both freezes and boils. Each such transition in nature is called a "phase transition" because the phase of the material changes—from solid to liquid to gas. It is fair to say that if we understand the phases and the conditions that govern the phase transitions for any material, we understand most of the essential physics.

Now, what makes this especially difficult is that in the region of a phase transition, matter appears as complicated as it ever gets. As water boils, turbulent eddies swirl, and bubbles explosively burst from the surface. However, in complexity there often lie the seeds of order. While a cow may seem hopelessly complex, we have seen how simple scaling arguments govern a remarkable number of its properties without requiring us to keep track of all the details. Similarly, we can never hope to describe specifically the formation of every bubble in a pot of boiling water. Nevertheless, we can characterize several generic features that are always present when, say, water boils at a certain temperature and pressure, and examine their scaling behavior.

For example, at normal atmospheric pressure, when water is at the boiling point, we can examine a small volume of the material chosen at random. We can ask ourselves the following question: Will this region be located inside a bubble, or inside water, or neither? On small scales, things are very complicated. For example, it is clear that it makes no sense to describe a single water molecule as a gas or a liquid. This is because it is the configuration of many, many molecules—how close together they are on average, for example—that distinguishes a gas from a liquid. It is also clear that it makes no sense to describe a small group of molecules moving around as being in a liquid or a vapor state. This is because as they move and collide, we can imagine that at times a significant fraction of the molecules are sufficiently far apart that they could be considered as being in the vapor state. At other times, they might be close enough together to be considered a liquid. Once we get to a certain size region, however, containing enough molecules, it becomes sensible to ask whether they are in the form of a liquid or a gas.

When water boils under normal conditions, both bubbles of water vapor and the liquid coexist. Thus water is said to undergo a "first-order" transition at the boiling point, 212° Fahrenheit at sea level. A macroscopic sample after sufficient time at exactly the boiling point will settle down and can be unambiguously characterized as being in either the liquid or the gaseous state. Both are possible at precisely this temperature. At a slightly lower temperature, the sample will always settle down in the liquid state; at a slightly higher temperature, it will settle down as vapor.

In spite of the intricate complexity of the local transitions that take place as water converts from liquid to gas at the boiling point, there is a characteristic volume scale, for a fixed pressure, at which it is sensible to ask which state the water is in. For all volumes smaller than this scale, local fluctuations in density are rapidly occurring that obscure the distinction between the liquid and gaseous states. For volumes larger than this, the average density fluctuates by a small enough amount so that this bulk sample has the properties of either a gas or a liquid.

It is perhaps surprising that such a complex system has such uniformity. It is a product of the fact that each drop of water contains an incredibly large number of molecules. While small groups of them may behave erratically, there are so many behaving in an average fashion that the few aberrant ones make no difference. It is similar to people, I suppose. Taken individually, everyone has his or her own reasons for voting for a political candidate. Some people even prepare write-in ballots for some local candidate of choice. There are enough of us, however, so that the TV stations, on the basis of exit polls, can predict quite quickly who will win. On average, all our differences cancel out.

Having discovered such hidden order, we can exploit it. We can, for instance, ask whether the scale at which the distinction between water and liquid becomes meaningful changes as we change the temperature and pressure combination at which water boils. As we increase the pressure, and thus decrease the difference between the density of water vapor and water liquid, the temperature at which water boils increases. If we now achieve this new temperature, we find, as you might expect after thinking about it, that because the difference in density between vapor and liquid is smaller, the size of the regions that regularly fluctuate between the two at the boiling point increases.

If we continue to increase the pressure, we find that at a certain conjunction of the pressure and temperature, called the *critical value,* the distinction between liquid and gas fails to have any meaning at all, even in an infinite sample. On all scales, fluctuations in density take place that are big enough to make it impossible to classify the bulk material as either liquid or gas. A little bit below this temperature, the density of the material is that appropriate to a liquid, and a little bit above the density is appropriate to a gas. But at this critical temperature, the water is neither or both, depending upon your point of view.

The specific configuration of water at the critical point is remarkable. On all scales, the material looks exactly the same. The material is "self-similar" as you increase the scale on which you examine it. If you were to take a high-magnification photograph of a small region, with differences in density appearing as differences in color, it would look the same as a photograph taken at normal magnification, with the same kind of variations but with the regions in question representing much bigger volumes. In fact, in water at the critical point, a phenomenon called *critical opulescence* occurs. Because the fluctuations in density are occurring at all scales, these fluctuations scatter light of all wavelengths, and suddenly the water no longer looks clear, but gets cloudy.

There is even something more fascinating about this state of water. Because it looks more or less exactly the same on all scales—

what is called *scale invariance*—the nature of the microscopic structure of the water (that is, the fact that water molecules are made of two hydrogen atoms and one oxygen atom) becomes irrelevant. All that characterizes the system at the critical point is the density. One could, for instance, mark $+1$'s for regions with a little density excess and -1's for regions with a deficit. For all intents and purposes, the water on any scale would then look schematically something like this:

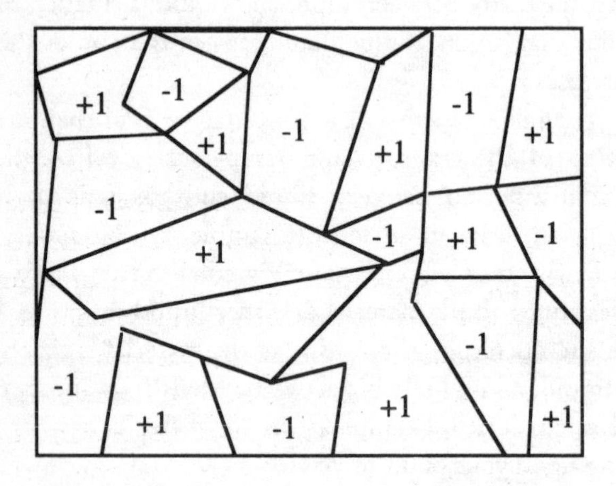

This is more than just a pictorial simplification. The fact that water can physically be distinguished *on all scales* only by this one effective degree of freedom, which can take on these two values, *completely determines* the nature of the phase transition in the region of this critical point. This means that the liquid-gas phase transition for water becomes *absolutely identical* to the phase transition in any other material, which at its own critical point can be described by a series of ± 1's.

For example, consider iron. Few people would confuse a block of iron with a glass of water. Now, as anyone who plays with magnets knows, iron can be magnetized in the presence of a magnet. Microscopically what happens is that each of the iron atoms is a little magnet, with a north and south pole. Under normal conditions, with no other magnets nearby, these iron atoms are randomly aligned, so that on average their individual magnetic fields cancel out to produce no macroscopic magnetic field. Under the influence of an external magnet, however, all the atomic magnets in the iron will line up in the same direction as the external field, producing a macroscopic iron magnet. If the external magnetic field points up, so will all the atomic magnets in the iron. If it points down, they too will point downward.

Now consider an idealized piece of iron, where the atomic magnets are constrained to point only up or down, but not in any other direction. At a low temperature, if an external magnetic field is present that starts out at some value pointing up, all the atomic magnets will be aligned in this direction. But if the external field decreases to zero, it will no longer dictate in which direction the atomic magnets can point. It turns out to remain energetically favorable on average for all of them to point in the same direction as one another, but the direction they choose is random. They could point either up or down. This means that such an iron magnet can have a phase transition. As the magnetic field from outside is tuned to zero, the little atomic magnets that had been following this field could now, due to some random thermal fluctuation, instead spontaneously align pointing down over the sample.

Mathematically, this begins to resemble the case of water. Replace "points up" with "density excess" and "points down" with "density deficit." Just as in the case of water, one finds that for such a magnet when there is no external magnetic field, there is a characteristic scale such that if one examines the sample on scales smaller than this, thermal fluctuations can still change the direction in which the magnets point. It will thus be impossible to state that the region has any net magnetic orientation. On scales larger than this, thermal fluctuations will not be able to cause the average magnetic orientation to change and it will remain constant. Furthermore, as one increases the temperature while keeping the external magnetic field zero, the sample will have a critical point. At this point, fluctuations in direction will persist throughout the entire sample in the same way on all scales, so that it will be useless to try to characterize the object by the orientation of its atomic magnets, even in an infinite sample.

What is important here is that, at the critical point, water and such a magnet are *exactly* the same. The fact that the actual microscopic structure of the two is totally different is irrelevant. Because the variations in the material at the critical point are characterized by just two degrees of freedom—up and down, or overdense and underdense—over all scales, even those much larger than the microscopic scales, the physics is insensitive to the microscopic differences. The behavior of water as it approaches its critical point, as labeled by whether it is a liquid or gas, is completely identical to that of the magnet, as labeled by whether its field points up or down. Any measurement you can make on one system will be identical for the other.

The fact that we can use the scaling properties of different systems, in this case their scale invariance near the critical point, to find uniformity and order in what is otherwise an incredibly complex situation is one of the great successes of what has become known as condensed matter physics. This approach, which has revolutionized the way we understand the physics of materials, was pioneered in the 1960s and 1970s by Michael Fisher and Kenneth Wilson at Cornell and Leo Kadanoff at the University of Chicago. The ideas developed in this endeavor have been imported throughout physics whenever complexity associated with scale has been an issue. Wilson was awarded the Nobel Prize in 1982 for his investigations of the applicability of these techniques to understanding the behavior not just of water but also of elementary particles, as I shall describe in the final chapter. What is important here is how they expose the underlying unity associated with the diverse and complex material world we deal with every day. It is not just the submicroscopic scales of elementary particles or the potentially infinite scales of cosmology where hidden connections can simplify reality. Think about it every time the teapot whistles, or the next time you wake up and see icicles on the window.

III

PRINCIPLES

5

The Search for Symmetry

"Is there any other point to which you would wish to draw my attention?"
"To the curious incident of the dog in the night-time"
"The dog did nothing in the night-time"
"That was the curious incident," remarked Sherlock Holmes.
—Sir Arthur Conan Doyle

WHEN AN artist thinks of symmetries, he or she may think of endless possibilities, of snowflakes, diamonds, or reflections in a pond. When a physicist thinks of symmetry, he or she thinks of endless impossibilities. What really drives physics is not the discovery of what happens but the discovery of what does not. The universe is a big place, and experience has taught us that whatever can happen does happen. What gives order to the universe is the fact that we can say with absolute precision that certain things never happen. Two stars may collide only once every million years per galaxy, which seems rare. Summed over all known galaxies, however, that is several thousand such events per year in the visible universe. Nevertheless you can wait 10 billion years and you will never see a ball on Earth fall *up.* That is order. Symmetry is the most important conceptual tool in modern physics precisely because it elucidates those things that do not change or cannot happen.

Symmetries in nature are responsible for guiding physicists in two important ways: They restrict the wealth of possibilities, and they fix the proper way to describe those that remain. What do we mean when we say something possesses a certain symmetry? Take a snowflake, for example. It may possess what a mathematician might call a sixfold symmetry. What this means is that you can hold the snowflake at any of six different angles, and it will look exactly the same. *Nothing has changed.* Now, let's take a more extreme but familiar example. Imagine a cow as a sphere! Why a sphere? Because it is the most symmetrical thing we can think of. Make any rotation, flip it in the mirror, turn it upside down, and it still looks the same. *Nothing has changed!* But what does this gain us? Well, because no rotation or flip can affect a sphere, the entire description of this object reduces to a single variable, its radius. Because of this, we were able to describe changes in its properties just by scaling this one variable. This feature is general: The more symmetrical something is, the fewer variables are needed to describe it completely.

I cannot overstress the importance of this one feature, and I shall praise it more later. For now, it is important to focus on how symmetries forbid change. One of the most striking things about the world, as Sherlock Holmes pointed out to the bewildered Watson, is that certain things do not happen. Balls do not spontaneously start bouncing up the stairs, nor do they pick up and roll down the hallway on their own. Vats of water don't spontaneously heat up, and a pendulum does not rise higher in the second cycle than it did in the first. All of these features arise from the symmetries of nature.

The recognition of this fact began in the work of the classical mathematical physicists of the eighteenth and nineteenth centuries, Joseph-Louis Lagrange in France and Sir William Rowan Hamilton, in England, who put Newton's mechanics on a more general, consistent mathematical footing. Their work reached fruition in the first half of this century through the brilliant German mathematician Emmy Noether. Unfortunately, her keen intellect did not help this woman in a man's world. Her untenured and unsalaried position at the distinguished mathematics department at Göttingen University was terminated by anti-Semitic laws in 1933—in spite of support from the greatest mathematician of the time, David Hilbert. (He argued unsuccessfully to the Göttingen faculty that they were part of a university and not a bathing establishment. Alas, university faculties have never been hotbeds of social awareness.)

In a theorem that bears her name, Noether demonstrated a mathematical result of profound importance for physics. After the fact, Noether's theorem seems eminently reasonable. Its formulation in physics goes essentially as follows: If the equations that govern the dynamical behavior of a physical system do not change when some transformation is made on the system, then for each such transformation there must exist some physical quantity that is itself *conserved,* meaning that it does not change *with time.*

This simple finding helps explain one of the most misunderstood concepts in popular science (including a number of undergraduate physics texts I have seen) because it helps show why certain things are impossible. For example, consider perpetual motion machines, the favorite invention of crackpot scientists. As I described chapter 1, they can be pretty sophisticated, and many reasonable people have been duped into investing in them.

Now, the standard reason why most machines of this type cannot work is the conservation of energy. Even without rigorously defining it, most people have a fairly good intuitive idea of what energy is, so that one can explain relatively easily why such a machine is impossible. Consider again the contraption illustrated on page 10. As I described there, after one complete cycle, each of its parts will have returned to its original position; if it was at rest at the beginning of the cycle, it would have to be at rest at the end. Otherwise it would have more energy at the end of the cycle than at the beginning. Energy would have had to be produced somewhere; since nothing has changed in the machine, no energy can be produced.

But the diehard inventor may say to me: "How do I know for sure that energy is conserved? What makes this law so special that it cannot be violated? All existing experiments may support this idea, but maybe there is a way around it. They thought Einstein was crazy too!"

There is some merit in this objection. We should not take anything on faith. So all these books tell undergraduates that Energy Is Conserved (they even capitalize it). And it is claimed that this is a universal law of nature, true for energy in all its forms. But while this is a very useful property of nature, the important issue is *Why?* Emmy Noether gave us the answer, and it disappoints me that many physics texts don't bother going this far. If you don't explain why such a wonderous quality exists, it encourages the notion that physics is based on some set of mystical rules laid down on high, which must be memorized and to which only the initiated have access.

So why *is* energy conserved? Noether's theorem tells us that it must be related to some symmetry of nature. And I remind you that a symmetry of nature tells us that if we make some transformation, everything still looks the same. Energy conservation is, in fact, related to the very symmetry that makes physics possible. We believe the laws of nature will be the same tomorrow as they are today. If they weren't, we would have to have a different physics text for every day of the week.

So we believe, and this is to some extent an assumption—but, as I shall show, a *testable* one—that *all* the laws of nature are invariant, that is, they remained unchanged, under time translation. This is a fancy way of saying that they are the same no matter when you measure them. But if we accept this for the moment, then we can show rigorously (that is, mathematically) that there must exist a quantity, which we can call energy, that is constant over time. Thus, as new laws of nature are discovered, we do not have to worry at each stage whether they will lead to some violation of the law of conservation of energy. All we have to assume is that the underlying physical principles don't change with time.

How, then, can we test our assumptions? First, we can check to see that energy is indeed conserved. This alone may not satisfy you or my inventor. There is another way, however. We can check the laws themselves over time to see that their predictions do not vary. This is sufficient to guarantee energy conservation. But aside from this new method of testing energy conservation, we have learned something much more important. We have learned what *giving up* energy conservation is tantamount to doing. If we choose not to believe that energy is conserved, then we must also believe that the laws of nature change with time.

It is perhaps not so silly to wonder whether, at least on some cosmic time scale, various laws of nature might actually evolve with time. After all, the universe itself is expanding and changing, and perhaps somehow the formulation of microphysical laws is tied to the macroscopic state of the universe. In fact, such an idea was proposed by Dirac in the 1930s. There are various large numbers that characterize the visible universe, such as its age, its size, the number of elementary particles in it, and so on. There are also some remarkably small numbers, such as the strength of gravity. Dirac suggested that perhaps the strength of gravity might vary as the universe expanded, getting weaker with time! He thought this might naturally explain why gravity might be so weak today, compared to the other forces in nature. The universe is old!

Since Dirac's proposal, many direct and indirect tests have been performed to check to see whether not only the strength of gravity but also the strength of the other forces in nature has been changing with time. To date, no positive evidence has been obtained. In fact, very stringent limits have been set on the variation of the fundamental constants over time. Observations of the abundance of light elements created in the big bang compared to theoretical predictions obtained by utilizing today's fundamental constants in the calculations are sufficiently good, for example, to imply that the strength of gravity could not have changed by more than about 20 percent in the 10 billion or so years since the universe was only one second old! Thus, as far as we can tell, gravity doesn't change with time.

Nevertheless, even if the formulation of microphysical laws *were* tied in some way to the macroscopic state of the universe, we would expect the underlying physical principles that tie the two together to remain fixed. In this case, it would always be possible to generalize our definition of energy so that it remains conserved. We are always free to generalize what we mean by energy as new physical principles arise on ever larger, or smaller, scales. But that *something,* which we can then call energy, remains conserved, as long as these principles do not change with time.

We have already had a number of occasions to revise our concept of energy. The most striking example involved Einstein's special and general theories of relativity. These, I remind you, imply that different observers may make different, but equally valid, measurements of fundamental quantities. As a result, these measurements must be considered relative to a specific observer rather than as absolute markers.

Now, when trying to understand the universe as a whole, or indeed any system where gravitational effects become strong, we must utilize a generalization of energy that is consistent with the curvature of space-time. However, if we consider the dynamics of the universe on scales that are small compared to the size of the visible universe, the effects of curvature become small. In this case, the appropriate definition of energy reduces to its traditional form. This in turn allows an example of how powerful the conservation of energy can be, even on cosmic scales. It is conservation of energy that determines the fate of the universe, as I now describe.

It is said that "what goes up must come down." Of course, like many an old saying, this one is incorrect. We know from our experience with spacecraft that it is possible to shoot something up off the Earth's surface so that it doesn't come back down. In fact, a universal velocity (the same for all objects) is required in order to escape the Earth's gravitational pull. (If this were not the case, the Apollo missions to the moon would have been much more difficult. The design of the spacecraft would have had to take explicitly into account how much each astronaut weighed, for example.) It is the conservation of energy that is responsible for the existence of a universal escape velocity.

We can define the energy of any object moving in the Earth's gravitational pull in terms of two parts. The first part depends upon the velocity of the object. The faster it is traveling, the more of this energy of motion—called *kinetic* energy, from the Greek word for motion—it has. Objects at rest have zero kinetic energy. The second part of the energy an object can have in a gravitational field is called *potential* energy. If a grand piano is hanging from a rope fifteen stories high, we know that it has a potential to do great damage. The higher something is, the more potential energy it has, because the greater might be the consequences should it fall.

The potential energy of well-separated objects is considered to be a negative quantity. This is merely a convention, but the logic behind it is this. An object at rest located infinitely far away from the Earth or any other massive body is defined to have zero total gravitational energy. Since the kinetic energy of such an object is zero, its potential energy must be zero as well. But since the potential energy of objects decreases as they get closer and closer to each other—as the piano does when it gets closer to the ground—this energy must get more negative as objects are brought together.

If we stick with this convention, then the two parts of the gravitational energy of any object in motion near the Earth's surface have opposite signs—one positive, one negative. We can then ask whether their sum is greater than or less than zero. This is the crucial issue. For *if energy is conserved,* then an object that starts out with negative total gravitational energy will never be able to escape without returning. Once it gets infinitely far away, even if it slows down to a halt, it would then have zero total energy, as I described above. This is, of course, greater than any negative value, and if the total energy starts out negative it can never become positive, or even zero, unless you add energy somehow. The velocity where the initial (positive) kinetic energy is exactly equal in magnitude to the (negative) potential energy, so that the total initial energy is zero, is the escape velocity. Such an object can, in principle, escape without returning. Since both forms of energy depend in the same way upon the mass of an object, the escape velocity is independent of mass. From the surface of the Earth, for example, the escape velocity is about 10 kilometers per second, or about 20,000 miles per hour.

If the universe is *isotropic*—the same everywhere—then whether or not the universe will expand forever is equivalent to whether or not an average set of well-separated galaxies will continue to move apart from each other indefinitely. And this is identical to the question of whether or not a ball thrown up from the Earth will come down. If the relative velocity of the galaxies, due to the background expansion, is large enough to overcome the negative potential energy due to their mutual attraction, they will continue to move apart. If their kinetic energy exactly balances their potential energy, the total gravitational energy will be zero. The galaxies will then continue to move apart forever, but will slow down over time, never quite stopping until they (or what remains of them) are infinitely far apart from one another. If you remember my characterization of a flat universe in which we think we live, this is how I described it. Thus, if we do live in a flat universe today, the total (gravitational) energy of the entire universe is zero. This is a particularly special value, and one of the reasons a flat universe is so fascinating.

Whether the universe is open or closed, whether it will end with a bang or a whimper, is determined by energy. The answer to one of the most profound questions of human existence—How will the universe end?—can be determined by merely measuring the expansion velocity of a large set of galaxies and their total mass, and comparing one with the other. If the total energy of such systems determined in this way is greater than or equal to zero, the universe will expand forever. The fate of the universe has become a matter of bookkeeping.

• • •

Another symmetry of nature goes hand in hand with time-translation invariance. Just as the laws of nature do not depend upon *when* you measure them, then they should not depend on *where* you measure them. As I have explained to some students' horror, if this were not so we would need an introductory physics class not just at every university but in every building!

The consequence of this symmetry in nature is the existence of a conserved quantity called *momentum,* which most of you are familiar with as inertia—the fact that things that have started moving tend to continue to move and things that are standing still stay that way. Conservation of momentum is the principle behind Galileo's observation that objects will continue to move at a constant velocity unless acted upon by some external force. Descartes called momentum the "quantity of motion," and suggested that it was fixed in the universe at the beginning, "given by God." We now understand that this assertion that momentum must be conserved is true precisely because the laws of physics do not change from place to place.

But this understanding was not always so clear. In fact, there was a time in the 1930s when it seemed that conservation of momentum, at the elementary-particle level, might have to be dispensed with. Here's why. Momentum conservation says that if a system is at rest and suddenly breaks apart into several pieces—such as when a bomb explodes—all the pieces cannot go flying off in the same direction. This is certainly intuitively clear, but momentum conservation makes it explicit by requiring that if the initial momentum is zero, as it is for a system at rest, it must remain zero as long as no external force is acting on the system. The only way that the momentum can be zero afterward is if, for every piece flying off in one direction, there are pieces flying off in the opposite direction. This is because momentum, unlike energy, is a directional quantity akin to velocity. Thus, nonzero momentum carried by one particle can be canceled only by a nonzero momentum in the opposite direction.

One of the elementary particles that make up the nucleus of atoms, the *neutron,* is unstable when it is isolated and will decay in about 10 minutes. It decays into a proton and an electron, which

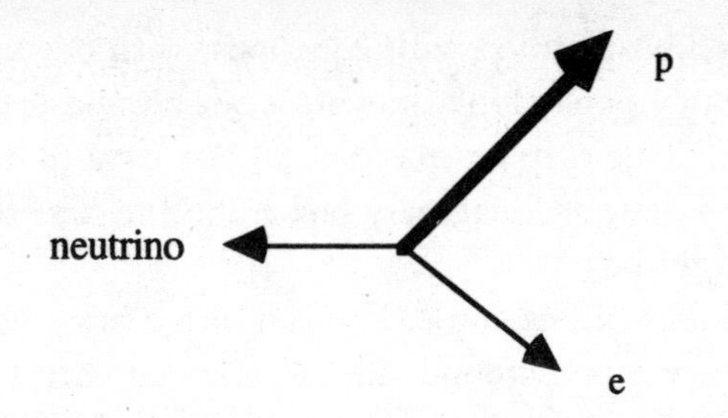

Now, inventing a hitherto undetected particle is not something to be taken lightly, but Pauli was not to be taken lightly, either. He had made important contributions to physics, notably the "Pauli exclusion principle," which governs the way electrons in atoms can behave. This Austrian-born genius also had an intimidating personality. He was famous for his habit of jumping up while listening to lectures and grabbing the chalk out of the speaker's hand if he thought the speaker was spouting nonsense. Moreover, the idea of giving up momentum and energy conservation, which had worked so well everywhere else in physics, seemed much more radical—in the spirit of creative plagiarism I discussed earlier—than what he was proposing. So the neutrino became an established part of physics long before it was experimentally observed in 1956, and before it became a standard part of astrophysics.

Today, of course, we would be more hesitant to give up momentum conservation, even at these small scales, because we recognize it as a consequence of a truly fundamental symmetry of nature. Unless we expect new dynamic laws of nature to depend somehow upon position, we can count on momentum conservation to be valid. And, of course, it doesn't apply only on subatomic scales. It is a fundamental part of understanding human-scale activities such as baseball, skating, driving, or typing. Whenever one finds an isolated system, with no external forces acting upon it, the momentum of this system is conserved, namely, it remains constant for all time.

Where does one fine such isolated systems? The answer is, anywhere you choose! There is a well-known cartoon that shows two scientists at a blackboard filled with equations, with one saying to the other: "Yes, but I don't think drawing a box around it makes it a Unified Theory" This may be true, but all you have to do to define a system is to draw an imaginary box around it. The trick lies in choosing the right box.

Consider the following example. You run into a brick wall with your car. Now draw a box around the car, and call that a system. Originally you were moving along at a constant speed; the momentum of the car was constant. Suddenly the wall comes along and stops you. Since your momentum changes to zero by the time you are at rest, the wall must exert a force on your system, the car. The wall will have to exert a certain force to stop you, depending upon your initial velocity.

Next, draw a box around the car and the wall. In this new system, no external forces are apparently at work. It would seem that the only thing acting on you is the wall, and the only thing acting on the wall is you. From this vantage point, what happens when you hit the wall? Well, if no external forces are acting in this system, momentum must be conserved—that is, must remain constant. Initially, you were moving along and had some momentum, and the wall was at rest, with zero momentum. After the crash, both you and the wall are apparently at rest. What happened to the momentum? It had to go somewhere. The fact that it seems to have disappeared here is merely a signal that the box is still not enough, namely, the system of you and the wall is not really isolated. The wall is anchored to the Earth. It is then clear that momentum can be conserved in this collision only if the Earth itself takes up the momentum that was initially carried by your car. The truly isolated system is thus made up of you, the wall, and the Earth. Since the earth is much more massive than your car, it doesn't have to move very much to take up this momentum, but nevertheless its motion must change! So, the next time someone tells you the Earth moved, rest assured that it did!

• • •

The search for symmetry is what drives physics. In fact, all the hidden realities discussed in the last chapter have to do with exposing new symmetries of the universe. Those that I have described related to energy and momentum conservation are what are called space-time symmetries, for the obvious reason that they have to do with those symmetries of nature associated with space and time, and to distinguish them from those that don't. They are thus integrally related to Einstein's Theory of Special Relativity. Because relativity puts time on an equal footing with space, it exposes a new symmetry between the two. It ties them together into a new single entity, space-time, which carries with it a set of symmetries not present if space and time are considered separately. Indeed, the invariance of the speed of light is itself a signal of a new symmetry in nature connecting space and time.

We have seen just how motion leaves the laws of physics invariant by providing a new connection between space and time. A certain four-dimensional space-time "length" remains invariant under uniform motion, just as a standard three-dimensional spatial length remains invariant under rotations. This symmetry of nature is possible only if space and time are tied together. Thus, purely spatial translations and purely time translations, which are themselves responsible for momentum and energy conservation, respectively, must be tied together. It is a consequence of special relativity that energy and momentum conservation are not separate phenomena. Together they are part of a single quantity called "energy-momentum." The conservation of this single quantity—which, in fact, requires some redefinition of both energy and momentum separately, as they are traditionally defined in the context of Newton's Laws—then becomes a single consequence of the invariances of a world in which space and time are tied together. In this sense, special relativity tells us something new: Space-time is such that we cannot have energy conservation without momentum conservation and vice versa.

There is one more space-time symmetry that I have so far alluded to only obliquely. It is related to the symmetry that results in energy-momentum conservation in special relativity, but it is much more familiar because it has to do with three dimensions and not four. It involves the symmetry of nature under rotations in space. I have described how different observers might see different facets of an object that has been rotated, but we know that fundamental quantities such as its total length remain unchanged under such a rotation. The invariance of physical laws as I rotate my laboratory to point in some different direction is a crucial symmetry of nature. We do not expect, for example, nature systematically to prefer some specific direction in space. All directions should be identical as far as the underlying laws are concerned.

The fact that physical laws are invariant under rotations implies that there is an associated quantity that is conserved. Momentum is related to the invariance of nature under spatial translations, while this new quantity is related to the invariance of nature under translations by an angle. For this reason, it is called *angular momentum.* Like momentum, angular momentum conservation plays an important role in processes ranging from atomic scales to human scales. For an isolated system, the angular momentum must be conserved. Indeed, for every example of momentum conservation, one can replace the words "distance" by "angle" and "velocity" by "angular velocity" to find some example of angular momentum conservation. It is a prime manifestation of creative plagiarism.

Here's one example. When my car hits another car that was at rest, and the bumpers lock so that the two move off together, the combination will move more slowly than my car alone was originally moving. This is a classic consequence of momentum conservation. The momentum of the combined system of the two cars must be the same after the collision as it was before. Since the combined mass of the system is larger than the mass of the original object that was moving, the combined object must move slower to conserve momentum.

On the other hand, consider a figure skater rotating very fast with her arms held tight by her side. When she spreads her arms outward, her rotation slows right down, as if by magic. This is a consequence of angular momentum conservation, just as the previous example was a consequence of momentum conservation. As far as rotations and angular velocities are concerned, an object with a bigger radius acts just like an object with a bigger mass. Thus, the act of raising her arms increases the radius of the skater's rotating body. Just as the two cars move together more slowly than one car as long as no external force acts on the system, so, too, a skater with increased radius will rotate more slowly than she will when her radius is smaller, as long as no external force is acting upon her. Alternatively, a skater who starts herself rotating slowly with her arms outstretched can then vastly increase her rotation speed by pulling her arms in. Thus are Olympic medals won.

There are other conserved quantities in nature that arise from symmetries other than those of space-time, such as electric charge. I will return to these later. For the moment, I want to continue with one strange facet of the rotational invariance of nature, which will allow me to introduce a ubiquitous aspect of symmetry that isn't always manifest. For example, even though the underlying laws of motion are rotationally invariant—that is, there is no preferred direction picked out by laws governing dynamics—the world isn't. If it were, then we should find it impossible to give directions to the grocery store. Left looks different than right; north is different than south; up is different than down.

It is easy for us to regard these as mere accidents of our circumstances, because that is exactly what they are. Were we somewhere else, the distinctions between left and right, and north and south, might be totally different. Nevertheless, the very fact that an accident of our circumstances can hide an underlying symmetry of the world is one of the most important ideas directing modern physics. To make progress, and to exploit the power of such symmetries, we have to look beneath the surface.

Many of the classic examples of hidden realities I discussed in the last chapter are related to this idea that symmetry can be masked. This idea goes by the intimidating name *spontaneous symme-*

try breaking, and we have already encountered it in a number of different guises.

A good example comes from the behavior of the microscopic magnets in a piece of iron, which I discussed at the end of the last chapter. At low temperature, when there is no external magnetic field applied to these magnets, it is energetically favorable for them all to line up in some direction, but the direction they choose is random. There is nothing in the underlying physics of electromagnetism that picks out one direction over another, and there is no prediction in advance that can precisely determine which direction they will choose. Once they have chosen, however, that direction becomes very special. An insect, sensitive to magnetic fields, living inside such a magnet would grow up assuming there was something intrinsically different about "north," that being the direction in which the microscopic magnets were aligned.

The trick of physics is to rise above the particular circumstances that may be attached to our own existence and attempt to peer beyond them. In every case I know of, this implies searching for the true symmetries of the world. In the case I just described, it would mean discovering that the equations governing the magnets were invariant under rotations and that north could be rotated to be south and the physics would still be the same.

The prototypical example of this is the unification of the weak and electromagnetic interactions. There, the underlying physics makes no distinction between the massless photon and the very massive Z particle. In fact, there is a symmetry of the underlying dynamics under which a Z can be turned into a photon and everything will look exactly the same. In the world in which we live, however, this same underlying physics has produced a specific realization, a solution of the equations—the "condensate" of particles occupying otherwise empty space—inside which the photon and the Z behave quite differently.

Mathematically, these results can be translated to read: A particular solution of a mathematical equation need not be invariant under the same set of transformations under which the underlying equation is invariant. Any specific realization of an underlying mathematical order, such as the realization we see when we look around the room, may break the associated underlying symmetry. Consider the example invented by the physicist Abdus Salam, one of those who won the Nobel Prize for his work on the unification of electromagnetism and the weak interaction: When you sit down at a circular dinner table, it is completely symmetrical. The wineglass on the right and the wineglass on the left are equivalent. Nothing but the laws of etiquette (which I can never remember) specifies which is yours. Once you choose a wineglass—say, the one on the right—everyone else's choice is fixed, that is, if everyone wants a wineglass. It is a universal fact of life that we live in one particular realization of what may be an infinity of possibilities. To paraphrase Rousseau: The world was born free, but it is everywhere in chains!

Why should we care so much about symmetries in nature, even those that are not manifest? Is it just some peculiar aesthetic pleasure that physicists derive, some sort of intellectual masturbation? Perhaps in part, but there is also another reason. Symmetries, *even* those that are not directly manifest, can completely determine the physical quantities that arise in the description of nature *and* the dynamical relations between them. In short, symmetry may *be* physics. In the final analysis, there may be nothing else.

Consider, for example, that energy and momentum—direct consequences of two space-time symmetries—together provide a description of motion that is completely equivalent to all of Newton's Laws describing the motion of objects in the gravitational field of Earth. All the dynamics—force producing acceleration, for example—follow from these two principles. Symmetry even determines the nature of the fundamental forces themselves, as I shall soon describe.

Symmetry determines for us what variables are necessary to describe the world. Once this is done, everything else is fixed. Take my favorite example again: a sphere. When I represent the cow as a sphere I am saying that the *only* physical processes that we need concern ourselves with will depend on the radius of the cow.

Anything that depends explicitly upon a specific angular location must be redundant, because all angular locations at a given radius are identical. The huge symmetry of the sphere has reduced a problem with a potentially large number of parameters to one with a single parameter, the radius.

We can turn this around. If we can isolate those variables that are essential to a proper description of some physical process, then, if we are clever, we may be able to work backward to guess what intrinsic symmetries are involved. These symmetries may in turn determine all the laws governing the process. In this sense we are once again following the lead of Galileo. Recall that he showed us that merely learning *how* things move is tantamount to learning *why* things move. The definitions of velocity and acceleration made it clear what is essential for determining the dynamical behavior of moving objects. We are merely going one step further when we assume that the laws governing such behavior are not merely made clear by isolating the relevant variables; rather, these variables alone *determine* everything else.

Let's return to Feynman's description of nature as a great chess game being played by the gods, which we are privileged to watch. The rules of the game are what we call fundamental physics, and understanding these rules is our goal. Feynman claimed that understanding these rules is all we can hope for when we claim to "understand" nature. I think now we can claim to go one step further. We suspect that these rules can be completely determined merely by exploring the configuration and the symmetries of the "board" and the "pieces" with which the game is played. Thus to understand nature, that is, to understand its rules, is equivalent to understanding its symmetries.

This is a very strong claim, and one done in rather grand generality. As I expect that you may be now both confused and skeptical, I want to go through a few examples to make things more explicit. In the process, I hope to give some idea of how physics at the frontier proceeds.

First, let me describe these ideas in the context of Feynman's analogy. A chess board is a rather symmetrical object. The pattern of the board repeats itself after one space in any direction. It is made of two colors, and if we interchange them, the pattern remains identical. Moreover, the fact that the board has 8 × 8 squares allows a natural separation into two halves, which can also be interchanged without altering the appearance of the board.

Now this alone is not sufficient to determine the game of chess because, for example, the game of checkers can also be played on such a board. However, when I add to this the fact that there are sixteen pieces on each of the two sides of a chess game, eight of which are identical, and of the others there are three sets of two identical pieces plus two loners, things become much more restricted. For example, it is natural to use the reflection symmetry of the board to lay out the three pieces with identical partners—the rook, the knight, and the bishop—in mirror-image pattern reflected about the center of the board. The two colors of the different opponents then replicate the duality of the board. Moreover, the set of moves of all the chess pieces is consistent with the simplest set of movements allowed by the layout of the board. Requiring a piece to move in only one color restricts the movement to diagonals, as is the case for the bishop. Requiring a pawn to capture a piece only when it is on the adjoining space of the same color requires also that captures be made on the diagonal, and so on. While I do not claim here that this is any rigorous proof that the game of chess is completely fixed by the symmetries of the board and the pieces, it is worth noting that there is only one variation of the game that has survived today. I expect that had there been other viable possibilities, they too would still be in fashion.

You might want to amuse yourself by asking the same question about your favorite sport. Would football be the same if it were not played on a 100-yard field divided into 10-yard sections? More important, how much do the rules depend upon the symmetries of the playing squad? What about baseball? The baseball diamond seems an essential part of the game. If there were five bases arranged on a pentagon, would you need four outs?

But why confine the discussion to sports? How much are the laws of a country determined by the configuration of legislators? And to ask a question that many people worried about military spending in the United States have asked: How much is defense planning dictated by the existence of four different armed forces: air force, army, navy, and marines?

To return to physics, I want to describe how it is that symmetries, even those that are not manifest, can manage to fix the form of known physical laws. I will begin with the one conservation law I have not yet discussed that plays an essential role in physics: the conservation of charge. All processes in nature appear to conserve charge—that is, if there is one net negative charge at the beginning of any process, then no matter how complicated this process is, there will be one net negative charge left at the end. In between, many charged particles may be created or destroyed, but only in pairs of positive and negative charges, so that the total charge at any time is equal to that at the beginning *and* that at the end.

We recognize, by Noether's theorem, that this universal conservation law is a consequence of a universal symmetry: We could transform all positive charges in the world into negative charges, and vice versa, and nothing about the world would change. This is really equivalent to saying that what we call positive and negative are arbitrary, and we merely follow convention when we call an electron negatively charged and a proton positively charged.

The symmetry responsible for charge conservation is, in fact, similar in spirit to a space-time symmetry that I have discussed in connection with general relatively. If, for example, we simultaneously changed all the rulers in the universe, so that what was 1 inch before might now read 2 inches, we would expect the laws of physics to look the same. Various fundamental constants would change in value to compensate for the change in scale, but otherwise nothing would alter. This is equivalent to the statement that we are free to use any system of units we want to describe physical processes. We may use miles and pounds in the United States, and every other developed country in the world may use kilometers and kilograms. Aside from the inconvenience of making the conversion, the laws of physics are the same in the United States as they are in the rest of the world.

But what if I choose to change the length of rulers by different amounts from point to point? What happens then? Well, Einstein told us that there is nothing wrong with the procedure. It merely implies that the laws governing the motion of particles in such a world will be equivalent to those resulting from the presence of some gravitational field.

What general relativity tells us, then, is that there is a general symmetry of nature that allows us to change the definition of length from point to point only if we also allow for the existence of such a thing as a gravitational field. In this case, we can compensate for the local changes in the length by introducing a gravitational field. Alternatively, we might be able to find a global description in which length remains constant from point to point, and then there need be no gravitational field present. This symmetry, called general-coordinate invariance, completely specifies the theory we call general relativity. It implies that the coordinate system we use to describe space and time is itself arbitrary, just like the units we use to describe distance are arbitary. There is a difference, however. Different coordinate systems may be equivalent, but if the conversion between them varies *locally*—that is, standard lengths vary from point to point—this conversion will also require the introduction of a gravitational field for certain observers in order for the predicted motion of bodies to remain the same. The point is this: In the weird world in which I choose to vary the definition of length from point to point, the trajectory of an object moving along under the actions of no other force will appear to be curved, and not straight. I earlier described this in my example of a plane traveling around the world as seen by someone looking at the projection on a flat map. I can account for this, and still be in agreement with Galileo's rules, only if I allow for an apparent force to be acting in this new frame. This force is gravity. The form of gravity, remarkably, can then be said to result from the general-coordinate invariance of nature.

This is not to imply that gravity is a figment of our imagination. General relativity tells us that mass *does* curve space. In this case, all coordinate systems we can choose will account for this curvature one way or another. It may be that locally one might be able to dispense with a gravitational field—that is, an observer falling freely will not "feel" any force on him- or herself, just as the astronauts orbiting Earth are freely falling and therefore feel no gravitational pull. However, the trajectories of different free-falling observers will bend with respect to one another—a sign that space is curved. We can choose whatever reference frame we wish. Fixed on Earth, we will experience a gravitational force. Freely falling observers may not. However, the dynamics of particles in both cases will reflect the underlying curvature of space, which *is* real and which is caused by the presence of matter. A gravitational field may be "fictional" in the sense that an appropriate choice of coordinates can get rid of it everywhere, but this is possible only if the underlying space in this case is completely flat, meaning that there is no matter around. One such example is a rotating coordinate system, such as you might find if you were standing against the wall in one of those carnival rides where a large cylindrical room turns and you get pushed to the outside. Inside the turning room, you might imagine that there is a gravitational field pulling you outward. There is, in fact, no mass that acts as a source of such a field, such as Earth does for our gravitational field. Those spectators watching recognize that what you may call gravitational field is merely an accident of a poor choice of coordinate system, one fixed to the rotating room. Curvature is real; a gravitational field is subjective.

I started out talking about electric charge and ended up talking about gravity. Now I want to do for electric charge what I did for length in space-time. Is there a symmetry of nature that *locally* allows me arbitrarily to choose my convention of the sign of electric charge and still keep the predictions of the laws of physics the same? The answer is yes, but only if there exists another field in nature acting on particles that can "compensate" for my local choice of charge in the same way that a gravitational field "compensates" for an arbitrarily varying choice of coordinate system.

The field that results from such a symmetry of nature is not the electromagnetic field itself, as you might imagine. Instead, this field plays the role of space-time curvature. It is always there if a charge is nearby, just as curvature of space is always there if a mass is around. It is not arbitrary. Instead, there is another field, related to the electromagnetic field, that plays a role analogous to the gravitational field. This field is called a *vector potential* in electromagnetism.

This weird symmetry of nature, which allows me locally to change my definition of charge or length at the expense of introducing extra forces, is called a gauge symmetry, and I referred briefly to it in an earlier chapter. Its presence, in different forms, in general relativity and electromagnetism was the reason Hermann Weyl introduced it and tried to unify the two. It turns out to be far more general, as we shall see. What I want to stress is that such a symmetry (1) *requires* the existence of various forces in nature, and (2) tells us what are those quantities that are truly "physical" and what are the quantities that are just artifacts of our particular reference frame. Just as the angular variables on a sphere are redundant if everything depends only on the radius of the sphere, so in some sense the gauge symmetry of nature tells us that electromagnetic fields and space-time curvature are physical and that gravitational fields and vector potentials are observer-dependent.

The exotic language of gauge symmetry would be mere mathematical pedantry if it were used only to describe things after the fact. After all, electromagnetism and gravity were understood well before gauge symmetry was ever proposed. What makes gauge symmetry important is its implications for the rest of physics. We have discovered in the last twenty-five years that all the known forces in nature result from gauge symmetries. This in turn has allowed us to build a new understanding of things we did not before understand. The search for a gauge symmetry associated with these forces has allowed physicists to distinguish the relevant physical quantities that underly these forces.

It is a general property of a gauge symmetry that there must exist some field which can act over long distances, associated with the ability to compensate for the freedom to vary the definition of certain properties of particles or space-time from point to point over long distances without changing the underlying physics. In the case of general relativity, this is manifested by the gravitational field; in electromagnetism, it is manifested by electric and magnetic fields (which themselves result from vector potentials). But the weak interaction between particles in nuclei acts only over very short distances. How can it be related to an underlying gauge symmetry of nature?

The answer is that this symmetry is "spontaneously broken." The same background density of particles in empty space that causes a Z particle to appear massive, while a photon, which transmits electromagnetism, remains massless, provides a background that can physically respond to the weak charge of an object. For this reason one is no longer free to vary locally what one means by positive and negative weak charge and thus the gauge symmetry which would otherwise be present is not manifest. It is as if there were a background electric field in the universe. In this case, there would be a big difference between positive and negative charges in that one kind of charge would be attracted by this field and another kind would be repelled by it. Thus, the distinction between positive and negative charge would no longer be arbitrary. This underlying symmetry of nature would now be hidden.

The remarkable thing is that spontaneously broken gauge symmetries are not entirely hidden. As I have described, the effect of a background "condensate" of particles in empty space is to make the W and Z particles appear heavy, while keeping the photon massless. The signature of a broken gauge theory is then the existence of heavy particles, which transmit forces that can act only over short distances—so-called short-range forces. The secret of discovering such an underlying symmetry that is broken is to look at the short-range forces and explore similarities with long-range forces—forces that can act over long distances, such as gravity and electromagnetism. This, in a heuristic sense at least, is exactly how the weak interaction was eventually "understood" to be a cousin of quantum electrodynamics, the quantum theory of electromagnetism.

Feynman and Murray Gell-Mann developed a phenomenological theory in which the weak interaction was put in the same form as electromagnetism in order to explore the consequences. Within a decade a theory unifying both electromagnetism and the weak interaction as gauge theories was written down. One of its central predictions was that there should be one part of the weak force that had never been previously observed. It would not involve interactions that mixed up the charge of particles, such as that which takes a neutral neutron and allows it to decay into a positive proton and a negative electron. Rather, it would allow interactions that would keep the charges of particles the same, just as the electric force can act between electrons without changing their charge. This "neutral interaction" was a fundamental prediction of the theory, which was finally unambiguously observed in the 1970s. This was perhaps the first case of a discovery of a symmetry that *predicted* the existence of a new force, rather than naming it in hindsight.

The weakness of the weak interaction is due to the fact that the gauge symmetry associated with it gets spontaneously broken. Thus—on length scales larger than the average distance between particles in the background condensate that affects the properties of the W and Z particles—these particles appear very heavy and the interactions mediated by them are suppressed. If other, new gauge symmetries were to exist in nature that got spontaneously broken at even smaller distance scales, the forces involved could easily be too weak to have yet been detected. Perhaps there are an infinite number of them. Perhaps not.

It then becomes relevant to ask whether all the forces in nature must result from gauge symmetries, even if they are spontaneously broken. Is there no other reason for a force to exist? We do not yet have a complete understanding of this issue, but we think the answer is likely to be that there is none. You see, all theories that do not involve such a symmetry are mathematically "sick," or internally inconsistent. Once quantum-mechanical effects are properly accounted for, it seems that an infinite number of physical parameters must be introduced in these theories to describe them properly. Any theory with an infinite number of parameters is no theory at all! A gauge symmetry acts to restrict the number of variables needed to describe physics, the same way a spherical symmetry acts to limit the number of variables needed to describe a cow. Thus, what seems to be required to keep various forces mathematically and physically healthy is the very symmetry that is responsible for their existence in the first place.

This is why particle physicists are obsessed with symmetry. At a fundamental level, symmetries not only describe the universe; they determine what is possible, that is, what *is* physics. The trend in spontaneous symmetry breaking has thus far always been the same. Symmetries broken at macroscopic scales can be manifest at smaller scales. As we have continued to explore ever smaller scales, the universe continues to appear more symmetrical. If one wishes to impose the human concepts of simplicity and beauty on nature, this must be its manifestation. Order *is* symmetry.

Once again, I have gotten carried away with phenomena at the high-energy frontier. There are also many examples of how symmetries govern the dynamical behavior of our everyday world and that are unrelated to the existence of new forces in nature. Let's return to these.

Until about 1950, the major place in which symmetry had manifested itself explicitly in physics was in the properties of materials such as crystals. Like Feynman's chess board, crystals involve a symmetrical pattern of atoms located on a rigid crystal lattice. It is the symmetrical pattern of these atoms that is reflected in the beautiful patterns of crystal surfaces such as those of diamonds and other precious stones. Of more direct relevance to physics, the movements of electronic charges inside a crystal lattice, like pawns in a chess game, can be completely determined by the symmetries of the lattice. For example, the fact that the lattice pattern repeats itself with a certain periodicity in space fixes the possible range of momenta of electrons moving inside the lattice. This is because the periodicity of the material inside the lattice implies that you can make a translation only up to a certain maximum distance before things look exactly the same, which is equivalent to not making any translation at all. I know that sounds a little like something you might read in Alice in Wonderland, but it does have a significant effect. Since momentum is related to the symmetry of physical laws under translations in space, restricting the effective size of space by this type of periodicity restricts the range of available momenta that particles can carry.

This single fact is responsible for the character of all of modern microelectronics. If I put electrons inside a crystal structure, they will be able to move freely about only within a certain range of momenta. This implies that they will have a fixed range of energies as well. Depending on the chemistry of the atoms and molecules in the crystal lattice, however, electrons in this energy range may be bound to individual atoms and not free to move about. Only in the case that this "band" of accessible momenta and energies corresponds to an energy range where electrons are free to flow does the material easily conduct electricity. In modern semiconductors such as silicon, one can, by adding a certain density of impurities that affect which energy range of electrons is bound to atoms, arrange for very sensitive changes in the conductivity of the materials to take place as external conditions vary.

Arguments such as these may in fact be relevant to the greatest mystery of modern condensed matter physics. Between 1911, when Onnes discovered superconductivity in mercury, and 1987, no material had ever been found that became superconducting at temperatures higher than 20° above absolute zero. Finding such a material had long been the holy grail of the subject. If something could be found that was superconducting at room temperature, for example, it might revolutionize technology. If resistance could be overcome completely without the necessity for complex refrigeration schemes, a whole new range of electrical devices would become practical. In 1987, as I mentioned earlier, two scientists working for IBM serendipitously discovered a material that became superconducting at 35° above absolute zero. Other similar materials were soon investigated, and to date materials that become superconducting at over 100° above absolute zero have been uncovered. This is still a far cry from room-temperature superconductivity, but it is above the boiling point of liquid nitrogen, which can be commercially produced relatively cheaply. If this new generation of "high-temperature" superconductors can be refined and manipulated into wires, we may be on the threshold of a whole new range of technology.

What is so surprising about these new superconductors is that they do not resemble, in any clear fashion, preexisting superconducting materials. In fact, in these materials, superconductivity requires *introducing* impurities into the material. Many of these materials are, in fact, *insulators* in their normal state—that is, they do not conduct electricity at all.

In spite of frantic efforts by thousands of physicists, no clear understanding of high-temperature superconductivity yet exists. But the first thing they focused on was the symmetry of the crystal lattice in these materials, which appears to have a well-defined order. Moreover, it is made of separate layers of atoms that appear to act independently. Current can flow along these two-dimensional layers, but not along the perpendicular direction. It remains to be seen whether these particular symmetries of high temperature superconductors are responsible for the form of the interactions that results in a macroscopic superconducting state of electrons, but if history is any guide, that is the best bet.

Whether or not such lattice symmetries revolutionize electrical technology, they have already played a role in revolutionizing biology. In 1905, Sir William Bragg and his son Sir Lawrence Bragg were awarded a Nobel Prize for a remarkable discovery. If X rays, whose wavelength is comparable to the distance between atoms in a regular crystal lattice, are shined at such materials, the scattered X rays form a regular pattern on a detecting screen. The nature of the pattern can be traced directly to the symmetry of the lattice. In this way, X-ray crystallography, as it has become known, has provided a powerful tool to explore the spatial configuration of atoms in materials and, with this, the structure of large molecular systems that may contain tens of thousands of regularly arrayed atoms. The most well known application of this technique is probably the X-ray crystallographic data interpreted by Watson and Crick and their colleagues, which led to the discovery of the double-helix pattern of DNA.

The physics of materials has not been confined to technological developments. It has provided perhaps the most intimate relationship known between symmetry and dynamics, via the modern understanding of phase transitions. I have already described how, near a certain critical value of some parameter such as temperature or magnetic fields, totally different materials can effectively act the same. This is because at the critical point, most detailed microphysics becomes irrelevant and symmetry takes over.

Water at its critical point and an iron magnet at its critical point behave the same because of two related reasons. First, I remind you that at this critical point, small fluctuations are occurring simultaneously on all scales so, for example, it is impossible to say on any scale whether one is observing water or water vapor. Since the material looks the same at all scales, local microphysical properties such as the particular atomic configuration of a water molecule must become irrelevant. Second, because of this, all that is necessary to characterize the configuration of water is its density: Is the region under consideration overdense or underdense? Water can be completely characterized by the same two numbers, +1 or −1, which we can use to identify the configuration of microscopic magnets inside iron.

Both of these crucial features are integrally related to the idea of symmetry. Water, and our iron magnets at their critical point, have become in some sense like our chessboard. There are two degrees of freedom that can be mapped into each other—black into white, overdense into underdense, up into down. It didn't have to be this way. The fundamental parameter describing the possible states of the system near its critical point could have, for example, had a larger set of possibilities, such as pointing anywhere around a circle, as microscopic magnets might in a material in which they were not constrained to point up and down. Such a material, at its critical point, might then look schematically like this:

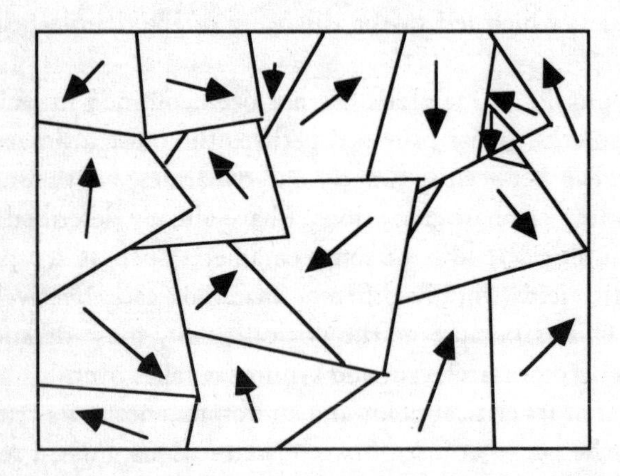

You might imagine that the fundamental characteristics of such a material as it approached its critical point would be different than water or the idealized iron magnets that were compared to water. And you would be correct. But what is the key difference between this picture and that shown on page 140 for water at its critical point? It is the set of possible values of the parameter that describe the transition—density, magnetic field direction, and so on. And what characterizes this set of possible values? The underlying symmetry of this "order parameter," describing the change in the "ordering" of the material. Can it take any value on a circle, a square, a line, a sphere?

Seen in this way, symmetry once again determines dynamics. The nature of a phase transition at the critical point is completely determined by the nature of the order parameter. But this order parameter is restricted by its symmetries. All materials with an order parameter having the same set of symmetries behave *identically* when they undergo a phase transition at the critical point. Once again, symmetry completely determines the physics.

This use of symmetry, in fact, allows us to make a powerful connection between the physics of materials and elementary-particle physics. For the above picture is nothing but an example of spontaneous symmetry breaking. The order parameter describing the direction of the local magnetic fields in the preceding picture can take any value on a circle. It possesses an intrinsic circular symmetry. Once it picks out some value, in some region, it "breaks" this symmetry by choosing one particular manifestation among all the possibilities. In the above example, at the critical point, this value continually fluctuates, no matter what scale you measure it over. However, away from the critical point the system will relax into one possible configuration on large enough scales, that is, liquid water, all magnets pointing up, all magnets pointing east, and so on. In elementary-particle physics, we describe the configuration of the ground state of the universe, the "vacuum," by the characteristics of any coherent configuration of elementary fields that have some fixed value in this state. The order parameters in this case are just the elementary fields themselves. If they relax to some value that is nonzero in otherwise empty space, then particles that interact with these fields will behave differently than particles that don't. The preexisting symmetry that might have existed among various elementary particles has been broken.

As a result, we expect that the spontaneous symmetry breaking that characterizes nature as we observe it today dies away on sufficiently small scales—where the order parameters, that is, the background fields, both fluctuate a great deal and, in any case, cannot alter the properties of particle motion on such small scales. Moreover, we also think spontaneous symmetry breaking disappeared at a very early stage in the big bang expansion. At that time, the universe was very hot. The same kind of phase transition that characterizes the liquefaction of water as temperature changes near the critical point can, in fact, occur for the ground state of the universe itself. At sufficiently high temperature, symmetries may become manifest because the order parameters, the elementary fields in nature, cannot relax to their low temperature values. And just as symmetries guide water through its phase transition, so, too, the symmetries of nature can guide the universe through its transitions. We believe that for every symmetry that is spontaneously broken at elementary scales today, there was at some sufficiently early time a "cosmic" phase transition associated with its breaking. Much of cosmology today is devoted to exploring the implications of such transitions, again governed by symmetry.

Returning to Earth, symmetry plays an even more powerful role in the phase transitions that govern the behavior of ordinary materials. We have seen that the symmetry of the order parameter of water, or magnets, or oatmeal, or whatever, can completely determine the behavior of these materials at their critical points. But perhaps the most powerful symmetry known in nature governs our very ability to describe these transitions. This symmetry, which has played a role from the beginning of this book, is scale invariance.

What is fundamental in being able to relate materials as diverse as magnets and water at their critical points is the fact that fluctuations at the critical point take place on all scales. The material becomes scale-invariant: It looks the same on all scales. This is a *very* special property, so special that it is not even shared by spherical cows! Recall that I was able to make my sweeping statements about the nature of biology by considering how spherical cows behave as you change their size. If the relevant physics remained scale-invariant, then cows of arbitrary magnitude would be allowed. But it doesn't, because the material of which cows are made does not change in density as you increase the size of the cow. This would be essential for quantities such as the pressure on the surface of a cow's belly to remain the same, and for the strength of a supercow's neck to grow in fixed proportion to its size.

But materials at the critical point of a phase transition *are* scale-invariant. The schematic diagrams of water and magnets such as I drew momentarily completely characterize the system on all scales. If I had used a microscope with higher resolving power, I would have seen the same kind of distribution of fluctuations. Because of this, only a very, very special kind of model will properly describe such a system near its critical point. The interesting mathematics of such models has become a focus for large numbers of both mathematicians and physicists in recent years. If one could, for example, categorize all possible models that possess scale invariance, then one could categorize all possible critical phenomena in nature. This, one of the most complex phenomena in nature, at least on a microscopic scale, could be completely predicted and thus, at least from a physics perspective, understood. Many of the people who are interested in scale invariance are, or were, particle physicists. This is because there is reason to believe, as I hope to make clear in the next chapter, that the ultimate Theory of Everything, if there is such a thing, may rest on scale invariance.

I want to end this chapter with a few remarks about where symmetry is taking us. This is, in fact, one of the few areas where one can get a glimpse of the origins of scientific progress at the frontier, when paradigms shift and new realities appear. This is where I can talk not just about things that are under control but also about things that are not. For the questions physicists ask about nature are often guided by symmetries we don't fully understand. Let me give you a few examples.

I have, throughout this chapter, made certain tacit assumptions about nature that seem unimpeachable. That nature should not care where and when we choose to describe it, for example, is the source of the two most important constraints on the physical world: energy and momentum conservation. In addition, while I can usually tell my right hand from my left, nature does not seem to care which is which. Would the physics of a world viewed in a mirror be any different? The sensible answer may seem to be no. However, our picture of what is sensible changed dramatically in 1956. In order to explain a puzzling phenomenon having to do with nuclear decays, two young Chinese-American theorists proposed the impossible: Perhaps nature herself can tell right from left! This proposal was quickly tested. The decay of a neutron, which produces an outgoing electron and a neutrino, inside a cobalt nucleus with its local magnetic field aligned in a certain direction, was observed. If left-right invariance were preserved, as many electrons should be emitted on average going off to the right as would go off to the left. Instead, the distribution was found to be asymmetrical. Parity, or left-right symmetry, was *not* a property of the weak interactions that governed this decay!

This came as a complete shock to the physics community. The two physicists, Chen Ning Yang, and Tsung Dao Lee, were awarded the Nobel Prize within a year of their prediction. *Parity violation,* as it became known, became an integral part of the theory of the weak interaction, and it is the reason that the neutrino, alone among particles in nature that feels only this interaction (as far as we know) has a very special property. Particles such as neutrinos, and also electrons, protons, and neutrons, behave as if they are "spinning," in the sense that in their interactions they act like little tops or gyroscopes. In the case of electrons and protons, which are charged, this spinning causes them to act like little magnets, with a north and south pole. Now, for an electron that is moving along, the direction of its internal magnetic field is essentially arbitrary. A neutrino, which is neutral, may not have an internal magnetic field, but its spin still points in some direction. It is a property of the parity violation of the weak interactions, however, that only neutrinos whose spin points along the same direction as their motion are emitted or absorbed during processes mediated by this interaction. We call such neutrinos left-handed, for no good reason except that this property is intimately related to the "handedness" observed in the distribution of particles emitted during nuclear decay.

We have no idea whether "right-handed" neutrinos exist in nature. If so, they need not interact via the weak interaction, so we might not know about them. But this does not mean they cannot exist. One can show, in fact, that if neutrinos are not exactly massless, like photons, it is highly probable that right-handed neutrinos might exist. If, indeed, any neutrino was found to have a nonzero mass, this would be a direct signal that some new physics, beyond the Standard Model, is required. It is for this reason that there is so much interest in the experiments now being performed to detect neutrinos emitted from the core of the sun. It has been shown that if the deficit observed in these neutrinos is real, then one of the most likely possible sources of such a deficit would be the existence of a nonzero neutrino mass. If true, we will have opened a whole new window on the world. Parity violation, which shocked the world but has now become a central part of our model of it, may point us in the right direction to look for ever more fundamental laws of nature.

Shortly after parity violation was discovered, another apparent symmetry of nature was found to be lacking. This is the symmetry between particles and their antiparticles. It had been previously thought, because antiparticles are identical in all ways to their particle partners except for, say, their electric charge, that if we replaced all the particles in the world by their antiparticles, the world would be identical. It is actually not this simple, because some particles with distinct antiparticles are electrically neutral, and can only be distinguished from their antiparticle partners through observing how each decays. In 1964, it was discovered that one such particle, called a neutral Kaon, decayed in a way that could not be reconciled with particle-antiparticle symmetry. Again, it appeared that the weak interaction was the culprit. The strong interaction between the quarks that make up Kaons had been independently measured to respect the symmetries of parity and particle-antiparticle interchange, to high precision.

However, in 1976, Gerard 't Hooft, in one of his many groundbreaking theoretical discoveries, demonstrated that what has become the accepted theory of the strong interaction, quantum chromodynamics, in fact *should* violate both parity and particle-antiparticle symmetry. Several clever theoretical proposals have been made to reconcile the apparent observed conservation of particle-antiparticleness in the strong interaction with 't Hooft's result. To date, we have no idea whether any of them are correct. Perhaps the most exciting involves the possible existence of new elementary particles, called *axions.* If these exist, they could easily be the dark matter that dominates the mass of the universe. Should they be detected as such, we will have made two profound discoveries. We will have learned some fundamental things about microscopic physics, as well as determining what the future evolution of the universe will be. If we do make such a discovery, symmetry considerations will have been the guiding light.

There are other symmetries of nature that exist, or do not exist, for reasons we don't understand. They form the fodder of modern theoretical research. Such problems prompt the major outstanding questions of elementary-particle physics: Why are there two other distinct sets, or "families" of elementary particles that resemble the familiar particles making up normal matter, except that these other particles are all much heavier? Why are the masses within each family different? Why are the "scales" of the weak interaction and gravity so different? The questions are framed in terms of symmetry. We expect, based on all our experience to date, that the answers will be as well.

6

It Ain't Over Till It's Over

We do not claim that the portrait we are making is the whole truth, only that it is a resemblance.
—Victor Hugo, *Les Misérables*

THERE IS a scene from a Woody Allen movie I particularly like in which a man obsessed with the meaning of life and death visits his parents, expresses confusion, and cries out for guidance. His father looks up and complains, "Don't ask me about the meaning of life. I don't even know how the toaster works!"

Throughout this book I too have stressed, perhaps not as cogently, the strong connection between the sometimes esoteric issues of interest at the cutting edge and in the physics of everyday phenomena. And so it seems appropriate to focus in this final chapter on how this connection is propelling us toward the discoveries-to-be in the twenty-first century. For the ideas I have discussed—going back to those that sprang forth from the small meeting in Shelter Island almost fifty years ago—have revolutionized the relationship between any possible future discoveries and our existing theories. The result has been perhaps the most profound, and unsung, realignment in our worldview that has taken place during the modern era. Whether or not one thinks there even exists such a thing as the Ultimate Answer still remains largely a matter of personal prejudice. However, modern physics has led us to the threshold of understanding why, at least directly, *it doesn't really matter.*

The central question I want to address here is: What guides our thinking about the future of physics, and why? I have spent the better part of this book describing how physicists have honed their tools to build our current understanding of the world, not least because it is precisely these tools that will guide our current approach to those things we have yet to understand. For this reason the discussion I am about to embark upon takes me full circle, back to approximation and scale. We thus will end where we began.

Physics has a future only to the extent that existing theory is incomplete. To get some insight into this, it is useful to ask what would be the properties of a complete physical theory if we had one. The simplest answer is almost tautological: A theory is complete if all the phenomena it was developed to predict are accurately predicted. But is such a theory necessarily "true," and, more important, is a true theory necessarily complete? For instance, is Newton's Law of Gravity true? It does predict with remarkable accuracy the motion of the planets around the sun and of the moon around the Earth. It can be used to weigh the sun to almost one part in a million. Moreover, Newton's Law is all that is necessary to compute the motion of projectiles near the Earth's surface to an accuracy of better than 1 part in 100 million. However, we now know that the bending of a light ray near the Earth is twice the amount that one might expect using Newton's Law. The correct prediction is obtained instead using general relativity, which generalizes Newton's Law and reduces to it in cases where the gravitational field is small. Thus, Newton's Universal Law of Gravity is incomplete. But is it untrue?

The preceding discussion may make the answer seem obvious. After all, one can measure deviations from Newton's Law. On the other hand, if every observation you are ever likely to make directly in your life is consistent with the predictions of Newton's Law, for all intents and purposes, it *is* true. To get around this technicality, suppose instead that one defines scientific truth to include only those ideas that are completely in accord with everything we know about the world. Newton's Law certainly does not meet this criterion. However, until at least the late nineteenth century, it did. Was it true then? Is scientific truth time-dependent?

You might say, especially if you are a lawyer, that my second definition, too, suffers from poor framing. I should remove the words "everything we know" and perhaps replace them with "everything that exists." The explanation is then watertight. But it is also useless! It becomes philosophy. It cannot be tested. We will never know whether we know everything that exists. All we can ever know is everything we know! This problem is, of course, insurmountable, but it has an important implication that is not often appreciated. It is a fundamental tenet of science that *we can never prove something to be true; we can only prove it to be false.* This is a very important idea, one that is at the basis of all scientific progress. Once we find an example in which a theory that may have worked correctly for millennia no longer agrees with observation, we then know that it must be supplemented—with either new data or a new theory. There is no arguing.

Nevertheless, there is a deeper and, I hope, less semantic issue buried here, and it is the one I want to concentrate on. What does it mean to say, even in principle, that any theory is *the* correct theory? Consider quantum electrodynamics (QED), the theory that reached completion as a result of the Shelter Island meeting in 1947. Some twenty years earlier, young Dirac had written down his relativistic equation for the quantum-mechanical motion of an electron. This equation, which correctly accounted for everything then known about electrons, presented problems, a number of which the Shelter Island meeting was convened to address, as I described. Nasty mathematical inconsistencies kept cropping up. The work of Feynman, Schwinger, and Tomonaga eventually presented a consistent method for handling these problems and producing meaningful predictions, which agreed completely with all the data. In the decades since the Shelter Island meeting, every measurement that has been made of the interactions of electrons and light has been in complete agreement with the predictions of this theory. In fact, it is the best-tested theory in the world. Theoretical calculations have been compared to ultrasensitive experimental measurements, and the agreement is now better than 9 decimal places in some cases! We could never hope for a more accurate theory than this.

Is QED, then, *the* theory of the interactions of electrons and photons? Of course not. We know, for instance, that if one considers processes at sufficiently high energies involving the heavy W and Z particles, then QED becomes part of a larger theory, the "electroweak" theory. At this stage, QED alone is not complete.

This is not a perverse accident. Even if the W and Z particles did not exist and electromagnetism was the only force we knew about in nature besides gravity, we could not call QED *the* theory of electrons and photons. Because what we have learned in the years following the Shelter Island meeting is that this statement, without further qualification, *makes no sense physically.* The incorporation of relativity and quantum mechanics, of which QED was the first successful example, has taught us that *every* theory like QED is meaningful only to the extent that we associate a dimensional *scale* with each prediction. For example, it is meaningful to say that QED *is* the theory of electron and photon interactions that take place at a distance of, say, 10^{-10} centimeters. On such a scale, the W and Z particles do not have a direct effect. This distinction may seem like nitpicking at the moment, but trust me, it isn't.

In chapter 1, I harped on the necessity of associating dimensions and scale with physical *measurements.* The recognition of the need to associate scales, of length or energy, with physical theory really began in earnest when Hans Bethe made the approximation that allowed him to calculate the Lamb shift five days after the Shelter Island meeting. I remind you that Bethe was able to turn an unmanageable calculation into a prediction using physical reasoning as a basis to ignore effects he did not understand.

Recall what Bethe was up against. Relativity and quantum mechanics imply that particles can spontaneously "pop" out of empty space only to disappear quickly, as long as they do so for too short a time to be directly measured. Nevertheless, the whole point of the Lamb shift calculation was to demonstrate that these particles *can* affect the measurable properties of ordinary particles, such as an electron in a hydrogen atom. The problem, however, was that the effects of all possible virtual particles, with arbitrarily high energies, appeared to make the calculation of the properties of the electron mathematically intractable. Bethe argued that somehow, if the theory were to be sensible, the effect of virtual particles of arbitrarily high energy acting over only very short time intervals, should be ignorable. He did not know at the time how to work with the complete theory, so he just threw out the effects of these high-energy virtual particles and hoped for the best. That is exactly what he got.

When Feynman, Schwinger, and Tomonaga figured out how to handle the complete theory, they found out that the effects of high-energy virtual particles were, indeed, consistently ignorable. The theory gave reasonable answers as any reasonable theory must. After all, if effects on extremely small time and distance scales compared to the atomic scales being measured were to be significant, there would be no hope of doing physics. It is like saying that in order to understand the motion of a baseball, one would have to follow in detail the forces acting at the molecular level during every millionth of a second of its travels.

It has been an implicit part of physics since Galileo that irrelevant information must be discarded, a fact I also stressed in chapter 1. This is true even in very precise calculations. Consider the baseball again. Even if we calculate its motion to the nearest millimeter, we are still making the assumption that we can treat it as a baseball. Actually, it is an amalgam of approximately 10^{24} atoms, each of which is performing many complicated vibrations and rotations during the flight of the ball. It is a fundamental property of Newton's Laws, however, that we can take an arbitrarily complicated object and divide its motion into two pieces: (1) the motion of the "center of mass," determined by averaging the position of all the individual masses under consideration, and (2) the motion of all the individual objects about the center of mass. Note that the center of mass need not be in a location where any mass actually exists. For example, the center of mass of a doughnut is right in the center, where the hole is! If we were to throw the doughnut in the air, it might twirl and spin in complicated ways, but the movement of the center of mass, the doughnut hole, would follow a simple parabolic motion first elucidated by Galileo.

Thus, when we study the motion of balls or doughnuts according to Newton's Laws, we are really studying what we now call an effective theory. A more complete theory must be a theory of quarks and electrons, or at least of atoms. But we are able to lump all these irrelevant degrees of freedom into something we call a ball—by which we mean the center of mass of a ball. The laws of motion of all macroscopic objects involve an effective theory of the motion of, and about, their center of masses. The effective theory of a ball's motion is all we need, and we can do so much with it that we tend to think of it as fundamental. What I will now argue is that all theories of nature, at least the ones that currently describe physics, are of necessity effective theories. Whenever you write one down, you are throwing things out.

The utility of effective theories was recognized in quantum mechanics early on. For example, in an atomic analogy to the center-of-mass motion of the ball I just discussed, one of the classic methods of understanding the behavior of molecules in quantum mechanics—which goes back to at least the 1920s—is to separate molecules into "fast" and "slow" degrees of freedom. Since the nuclei in molecules are very heavy, their response to molecular forces will involve a smaller, and slower, variation than, say, the electrons speedily orbiting them. Thus one might follow a procedure such as this to predict their properties. First, imagine the nuclei are fixed and unchanging and then calculate the motion of the electrons about these fixed objects. Then, as long as the nuclei are slowly moving, one would not expect this motion significantly to affect the electrons' configuration. The combined set of electrons will just smoothly track the motion of the nuclei, which will in turn be affected only by the average electron configuration. The effect of the individual electrons thus "decouples" from the motion of the nuclei. One can then describe an effective theory of the nuclear motion keeping track only of the nuclear degrees of freedom explicitly and replacing all the individual electrons by some single quantity representing the average charge configuration. This classic approximation in quantum mechanics is called the Born-Oppenheimer theory, after the two well-known physicists who first proposed it, Max Born and Robert Oppenheimer. This is just like describing the motion of the ball by merely keeping track of the center of mass of the ball, plus perhaps also the collective motion of all the atoms about the center of mass—namely, the way the ball spins.

Take another, more recent example, related to superconductivity. I have described how, in a superconductor, electron pairs bind together into a coherent configuration. In such a state, one need not describe the material by keeping track of all the electrons individually. Because it takes so much energy to cause an individual electron to deviate from the collective pattern, one can effectively ignore the individual particles. Instead, one can build an effective theory in terms of just a single quantity describing the coherent configuration. This theory, proposed by London in the 1930s and further developed by the Soviet physicists Landau and Ginsberg in 1950, correctly reproduces all the major macroscopic features of superconducting materials—including the all-important Meissner effect, which causes photons to behave like massive objects inside superconductors.

I have already pointed out that separating a problem into relevant and irrelevant variables is not itself a new technique. What the integration of quantum mechanics and relativity has done, however, is to require the elimination of irrelevant variables. In order to calculate the results of any measurable microscopic physical process, we must ignore not just a few, but an *infinite* number of quantities. Thankfully, the procedure begun by Feynman and others has demonstrated that these can be ignored with impunity.

Let me try to describe this key point in a more concrete context. Consider the "collision" of two electrons. Classical electromagnetism tells us that the electrons will repel each other. If the electrons are initially moving very slowly, they will never get close together, and classical arguments may be all that is necessary to determine correctly their final behavior. But if they are moving fast enough initially to get close together, on an atomic scale, quantum-mechanical arguments become essential.

What does an electron "see" when it reacts to the electric field of another electron? Because of the existence of virtual pairs of particles and antiparticles burping out of the vacuum, each electron carries a lot of baggage. Those positive particles that momentarily pop out of the vacuum will be attracted to the electron, while their negative partners will be repelled. Thus, the electron in some sense carries a "cloud" of virtual particles around with it. Since most of these particles pop out of the vacuum for an extremely short time and travel an extremely small distance, this cloud is for the most part pretty small. At large distances we can lump together the effect of all the virtual particles by simply "measuring" the charge on an electron. In so doing, we are then lumping into a single number the potentially complicated facets of the electric field due to each of the virtual particles that may surround an electron. This "defines" the charge on an electron that we see written down in the textbooks. This is the *effective* charge that we measure at large distances in an apparatus in the laboratory, by examining the motion of an electron in, say, a TV set, when an external field is applied.

Thus, the charge on the electron is a fundamental quantity only to the extent that it describes the electron as measured *on a certain scale!* If we send another electron closer to the first electron, it can spend time inside the outskirts of this cloud of virtual particles and effectively probe a different charge inside. In principle, this is an example of the same kind of effect as the Lamb shift. Virtual particles can affect the measured properties of real particles. What is essential here is that they affect properties such as the electron's charge differently depending upon the scale at which you measure it.

If we ask questions appropriate to experiments performed on a certain length scale or larger, involving electrons moving with energies that are smaller than a certain amount, then we can write down a complete effective theory that will predict every measurement. This theory will be QED with the appropriate free parameters, the charge on the electron, and so on, now *fixed* to be those appropriate to the scale of the experiment, as determined by the results of experiment. All such calculations, however, of necessity effectively discard an infinite amount of information—that is, the information about virtual processes acting on scales smaller than our measurements can probe.

It may seem like a miracle that we can be so blasé about throwing out so much information, and for a while it seemed that way even to the physicists who invented the procedure. But upon reflection, if physics is to be possible at all, it has to work out this way. After all, the information we are discarding need not be correct! Every measurement of the world involves some scale of length or energy. Our theories too are *defined* by the scale of the physical phenomena we can probe. These theories may predict an infinite range of things at scales beyond our reach at any time, but why should we believe any of them until we measure them? It would be remarkable if a theory designed to explain the interactions of electrons with light were to be absolutely correct in all its predictions, down to scales that are orders and orders of magnitude smaller than anything we currently know about. This might even be the case, but either way, should we expect the correctness of the theory at the scales we can now probe to be held hostage by its potential ability to explain everything on smaller scales? Certainly not. But in this case, all the exotic processes predicted by the theory to occur on much smaller scales than we can now probe had better be irrelevant to its predictions for the comparison with present experiments, precisely because we have every reason to believe that these exotic processes could easily be imaginary remnants of pushing a theory beyond its domain of applicability. If a theory had to be correct on all scales to answer a question about what happens at some fixed scale, we would have to know The Theory of Everything before we could ever develop A Theory of Something.

Faced with this situation, how can we know whether a theory is "fundamental"—that is, whether it has a hope of being true on all scales? Well, we can't. All physical theories we know of must be viewed as effective theories precisely because we have to ignore the effects of new possible quantum phenomena that might occur on very small scales in order to perform calculations to determine what the theory predicts on larger, currently measurable, scales.

But as is often the case, this apparent deficiency is really a blessing. Just as we could predict, at the very beginning of this book, what the properties of a supercow should be by scaling up from the properties of known cows, so the fact that our physical laws are scale-dependent suggests that we might be able to predict how they, too, evolve as we explore ever smaller scales in nature. The physics of today then can give a clear signpost for the physics of tomorrow! In fact, we can even predict in advance when a new discovery is *required.*

Whenever a physical theory either predicts nonsense or else becomes mathematically unmanageable as the effects of smaller and smaller scale virtual quantum-mechanical processes are taken into account, we believe that some new physical processes must enter in at some scale to "cure" this behavior. The development of the modern theory of the weak interaction is a case in point. Enrico Fermi wrote down in 1934 a theory describing the "beta" decay of the neutron into a proton, electron, and neutrino—the prototypical weak decay. Fermi's theory was based on experiment and agreed with all the known data. However, the "effective" interaction that was written down to account for a neutron decaying into three other particles was otherwise ad hoc, in that it was not based on any other underlying physical principles beyond its agreement with experiment.

Once quantum electrodynamics had been understood, it soon became clear that Fermi's weak interaction differed fundamentally in nature from QED. When one went beyond simple beta decay, to explore what the theory might predict to occur at smaller scales, one encountered problems. Virtual processes that could take place at scales hundreds of times smaller than the size of a neutron would render the predictions of the theory unmanageable once one tried to predict the results of possible experiments at such scales.

This was not an immediate problem, since no experiments would be directly capable of exploring such scales for over fifty years after Fermi invented his model. Nevertheless, well before this, theorists began to explore possible ways to extend Fermi's model to cure its illnesses. The first step to take around this problem was clear. One could calculate the distance scale at which problems for the predictions of the theory would begin to be severe. This scale corresponded to about 100 times smaller than the size of a neutron—much smaller than that accessible at any then-existing facility. The simplest way to cure the problem was then to suppose that some new physical processes, not predicted in the context of Fermi's theory alone, could become significant at this scale (and not larger), and could somehow counter the bad behavior of the virtual processes in Fermi's theory. The most direct possibility was to introduce new virtual particles with masses about 100 times the mass of the neutron, which would make the theory better behaved. Because these particles were so massive, they could be produced as virtual particles only for very short times, and could thus move over only very small distance scales. This new theory would give identical results to Fermi's theory, as long as experiments were performed at scales that would not probe the structure of the interaction, that is, at scales greater than the distance traveled by the massive virtual particles.

We have seen that positrons, which were predicted to exist as part of virtual pairs of particles in QED, also exist as real measurable particles, if only you have enough energy to create them. So too for the new superheavy virtual particles predicted to exist in order to cure Fermi's theory. And it is these superheavy W and Z particles that were finally directly detected as real objects in a very high energy particle accelerator built in Geneva in 1984, about twenty-five years after they were first proposed on theoretical grounds.

As I have described, the W and Z particles make up part of what is now known as the Standard Model of particle physics describing the three nongravitational forces in nature: the strong, weak, and electromagnetic forces. This theory is a candidate "fundamental" theory, in that nothing about the possible virtual processes at very small length scales that are predicted to occur in the theory directly requires new processes beyond those predicted in the theory at these scales. Thus, while no one actually believes it, this theory could in this sense be complete. By the same token, nothing precludes the existence of new physics operating at very small length scales. Indeed, there are other strong theoretical arguments that suggest that this is the case, as I shall describe.

While theories like Fermi's, which are "sick," give clear evidence for the need for new physics, theories like the Standard Model, which are not, can also do so simply because their formulation is scale-dependent—namely, they depend intrinsically on what scale we perform the experiments to measure their fundamental parameters. As processes involving virtual particles acting on ever smaller scales are incorporated in order to compare with the results of ever more sensitive experiments, the value of these parameters is predicted to change, in a predictable way! For this reason, the properties of the electron that participates in atomic process on the atomic scales are not exactly the same as those of an electron that interacts with the nucleus of an atom on much smaller, nuclear scales. But most important, the difference is calculable!

This is a remarkable result. While we must give up the idea that the Standard Model is a single, inviolable theory, appropriate at all scales, we gain a continuum of effective theories, each appropriate at a different scale and all connected in a calculable way. For a well-behaved theory like the Standard Model, then, we can literally determine how the laws of physics should change with scale!

This remarkable insight about the scale-dependence of physics as I have described it is largely the work of Ken Wilson, dating only from the 1960s and for which he was awarded the Nobel Prize. It originated as much from the study of the physics of materials as it did from particle physics. Recall that the scaling behavior of materials is the crucial feature that allows us to determine their properties near a phase transition. For example, my discussion of what happens when water boils was based on how the description of the material changed as we changed the scale on which we observed it. When water is in its liquid form, we may, if we observe on very small scales, detect fluctuations that locally make its density equivalent to its value in a gaseous state. However, averaged over ever larger and larger scales, the density settles down to be its liquid value once we reach some characteristic scale. What are we doing when we average over larger scales? We average out the effect of the smaller-scale phenomena, whose details we can ignore if all we are interested in is the macroscopic properties of liquid water. However, if we have a fundamental theory of water—one that can incorporate the small scale behavior—we can try to calculate exactly how our observations should vary with scale as the effects of fluctuations on smaller scales are incorporated. In this way, one can calculate all the properties of materials near a critical point where, as I have said, the scaling behavior of the material becomes all important. The same techniques applied to normal materials apply to the description of the fundamental forces in nature. Theories like QED contain the seeds of their own scale dependence.

Considerations of scale open up a whole new world of physics in just the same way as they did when we first encountered them in the context of our spherical cow at the beginning of this book. Indeed, returning to that example, we can see explicitly how this happens. First, if I determine by experiment the density and the strength of a normal cow's skin, I can tell you what will be the density of a supercow, which is twice as big. Moreover, I can predict the results of any measurement performed on such a supercow.

Is then the Spherical Cow Theory an Ultimate Theory of Cows? *A priori* we could never prove that it is, but there are three different ways of finding out if it isn't or is unlikely to be: (1) at a certain scale the theory itself predicts nonsense, (2) the theory strongly suggests that something even more simple than a sphere can do the same job, or (3) we can perform an experiment at some scale that distinguishes features not predicted in the theory. Here is such an experiment. Say I throw a chunk of salt at a spherical cow. The prediction is that it will bounce off:

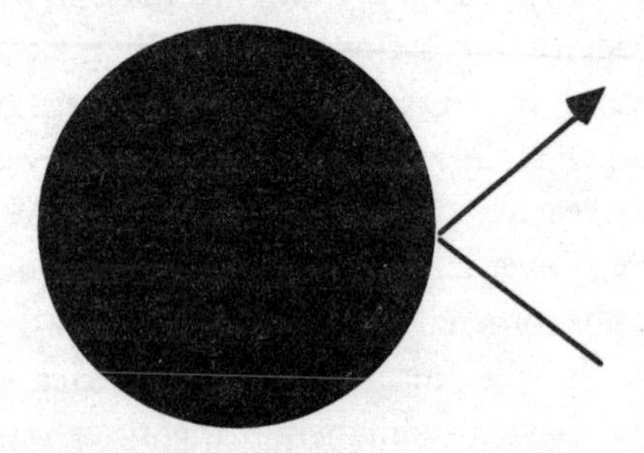

In reality, if I throw a chunk of salt in one direction at a cow, it will not bounce off. I will discover a feature not accounted for in the original theory—a hole, where the mouth is.

By the same token, exploring the scale dependence of the laws of nature provides crucial mechanisms to hunt for new fundamental physics. My brief history of the weak interaction was one classic example. I'll now describe some other, more current ones.

The scaling laws of fundamental physics can follow either a trickle-down or a bottom-up approach. Unlike economics, both techniques work in physics! We can explore those theories we understand at accessible length scales and see how they evolve as we decrease the scale in an effort to gain new insights. Alternatively, we can invent theories that may be relevant at length scales much smaller than we can now probe in the lab and scale them up, systematically averaging over small scale processes, to see what they might predict about physical processes at the scales we can now measure.

These two approaches encompass the span of research at the frontier today. I described in chapter 2 how the theory of the strong interaction, which binds quarks inside protons and neutrons, was discovered. The idea of asymptotic freedom played an essential role. Quantum chromodynamics (QCD), the theory of the strong interaction, differs from QED in one fundamental respect: the effect of virtual particles at small scales on the evolution of the parameters of the theory is different. In QED, the effect of the virtual-particle cloud that surrounds the electron is to "shield" to some extent its electric charge from the distant observer. Thus, if we probe very close to an electron, we will find the charge we measure effectively increases compared to the value we would measure if we probed it from the other side of the room. On the other hand, and this was the surprise that Gross, Wilczek, and Politzer discovered, QCD (and *only* a theory like QCD) can behave in just the opposite way. As you examine the interactions of quarks that get closer and closer together, the effective strong charge they feel gets weaker. Each of their clouds of virtual particles effectively increases their interaction with distant observers. As you probe further inside this cloud, the strength of the strong interaction among quarks gets weaker!

Furthermore, armed with a theory that correctly describes the small distance interactions of quarks, we can try to see how things change with scale, as we increase the scale. By the time you get to the size of the proton and neutron, one might hope to be able to average over all the individual quarks and arrive at an effective theory of just protons and neutrons. Because by this scale the interactions of quarks is so strong, however, no one has yet been able to carry it out directly, although large computers are being devoted to the task.

The great success of scaling arguments applied to the strong interaction in the early 1970s emboldened physicists to turn them around, to look to scales smaller than those which can be probed using the energy available in current laboratories. In this sense, they were following the lead of Lev Landau, the Soviet Feynman. By the 1950s this brilliant physicist had already demonstrated the fact that electric charge on electrons effectively increases as you reduce the distance scale at which you probe the electron. In fact, he showed that, at an unimaginably small scale, if the processes continued as QED predicted, the effective electric charge on the electron would become extremely large. This was probably the first signal, although it was not seen as such at the time, that QED as an isolated theory needed alteration before such small scales were reached.

QED gets stronger as the scale of energy is increased, and QCD gets weaker. The weak interaction strength is right in the middle. Round about 1975, Howard Georgi, Helen Quinn, and Steven Weinberg performed a calculation that altered our perception of the high-energy frontier. They explored the scaling behavior of the strong, weak, and electromagnetic interactions, under various assumptions about what kinds of new physics might enter as the scale of energy increased, and found a remarkable result. It was quite plausible that, at a scale roughly fifteen orders of magnitude smaller in distance than had ever been probed in the laboratory, the strength of all three fundamental interactions could become identical. This is exactly what one would expect if some new symmetry might become manifest at this scale which would relate all these interactions. The notion that the universe should appear more symmetrical as we probe smaller and smaller scales fit in perfectly with this discovery. The era of Grand Unified Theories, in which all the interactions of nature, aside from gravity, arise from a single interaction at sufficiently small scales, had begun.

It is almost twenty years later and still we have no further direct evidence that this incredible extrapolation in scale is correct. Recent precision measurements of the strength of all the forces at existing laboratory facilities have, however, given further support for the possibility that they could all become identical at a single scale. Whether or not this idea is correct, this result, more than any other in the postwar era, turned theoretical and experimental physicists' minds toward exploring the possibility of new physics on scales vastly different than those we could yet measure directly in the lab. The consequences have been mixed. The previous close connection between theory and experiment, which has always governed the progress of particle physics and indeed physics itself, has diminished. On the other hand, the stakes have increased. Some physicists now talk of a Theory of Everything.

There is, in physics, one astronomically high-energy scale that has been staring us in the face for the better part of this century. Fermi's weak interaction theory is not the only fundamental theory that is clearly sick at high energies and small distances. General relativity is another. When one tries to incorporate quantum mechanics and gravity, numerous problems arise. Foremost among them is the fact that at a scale roughly nineteen orders of magnitude smaller than the size of a proton, the effects of virtual particles in gravitational interactions become unmanageable. Like Fermi's theory, gravity is not a theory that can be fundamental as it stands. Some new physics must play a role to alter the behavior of the theory at these small scales.

One of the most remarkable of all of the possibilities that have been proposed to date, and many people's best bet, involves a bold new suggestion that perhaps new fundamental physics will one day end after all. If, at the scale where "quantum gravity"—as the merger of general relativity and quantum mechanics is called—becomes sick, a brand-new kind of physical theory emerges, based on a new kind of mathematics that pushes the limits of our existing knowledge, it has been argued that yet new symmetries might become manifest that cause the scale dependence of physical theory to stop. If this is the case, then the theory defined at this point can be truly called "complete." It would require no variation of its parameters as processes on ever smaller and smaller scales are examined. In principle, every result of every experiment that could ever be performed, on any scale at all, could be predicted by a single such theory. On large scales, any such theory would have to be able to be shown to reduce to the effective theories we now call the Standard Model, plus gravity. On small scales, it would display its true, unaltered form. It might even address such profound issues as the one which apparently most interested Einstein: was there any choice in the creation of the universe?

This is a remarkable dream, but at this point that is all it is. Research in this area would now be almost pure mathematics at this point were it not for the fact that the kind of theories that are needed to so to describe the physical world on very small scales—theories that become independent of scale—have been discussed in the context of other areas of physics—areas where we can perform experiments in a kitchen! After all, I described how normal materials such as boiling water behave in a very special way at their critical point—their properties become independent of scale. Theorists working to describe water and other materials at their critical points have been helping develop the necessary mathematical tools to describe such "scale-invariant" physics, which one day may explain not just the behavior of water but a theory of everything else under the sun as well.

Speculation about a Theory of Everything is interesting, but I don't want to end this book with it. Rather, I want to return to the equally likely possibility that the quest for Universal Truth could instead be misconceived. There may be an infinity of physical laws left to discover as we continue to probe the extremes of scale. No matter. We have learned that we can, and at present must, perform physics in a world of effective theories, which insulate the phenomena we understand from those that we have yet to discover. Scientific truth no longer requires the expectation that the theories we work with have to be truly fundamental. In this sense, physics is still clearly guided by the same principles introduced by Galileo 400 years ago, and indeed the very same principles I introduced at the beginning and throughout this book. Our present "fundamental" theories all involve intrinsic irremovable approximations, but we can use them with impunity. We are guided by ignoring the irrelevant. What is irrelevant is now, as ever, determined by the dimensional nature of physical quantities. These determine the scale of the problems we face, and thus also those we can safely ignore. All the while we creatively adapt old ideas to new situations. In so doing, we have been motivated to reach beyond our limited, human vantage point to glimpse what so far has always been a simpler, and more symmetric, reality beyond. Everywhere we look we still see spherical cows!

Notes

1. Galileo Galilei, *Dialogues Concerning Two New Sciences,* trans. Henry Crew and Alfonso de Salvio (New York: Dover, 1954; original ed., 1914), pp. 67, 64.
2. James Clerk Maxwell, *The Scientific Papers of James Clerk Maxwell,* ed. W. D. Niven (New York: Dover, 1965).
3. *Discoveries and Opinions of Galileo,* trans. Stillman Drake (New York: Anchor Books, 1990).
4. Richard Feynman, *The Character of Physical Law* (Cambridge, Mass.: MIT Press, 1965).
5. Robertson Davies, *Fifth Business* (Toronto: Macmillan, 1970).
6. Richard P. Feynman, Robert B. Layton, and Matthew Sands, *The Feynman Lectures on Physics,* vol. 2 (Reading, Mass.: Addison-Wesley, 1965).
7. Galilei, *Dialogues Concerning Two New Sciences,* p. 153.
8. Ibid., p. 166.
9. Feynman, *The Character of Physical Law.*
10. Ludwig Wittgenstein, *Tractatus Logico-Philosophicus* (London: Routledge and Kegan Paul, 1958).

INDEX